Ask More, Learn More

Ask More, Learn More

Creating a Culture of Curiosity

VICTORIA OLIVADOTI

ASK MORE, *LEARN MORE*
Creating a Culture of Curiosity

Contents

What's Inside?

Why does ***Questioning More Lead to Learning More?***

Why is questioning essential to spark curiosity?

How can we **Help** students overcome
their fear of asking questions?

How does Neuroscience support the need to ask questions?

When and **How** should students ask questions?

Why is questioning a necessary skill in this
rapidly changing digital world?

Why do Artificial Apps call for everyone to ask questions?

Why is it crucial that parents of college students
teach their children to be question askers?

Why is an idle brain necessary for
finding answers to questions?

Why is it time to create a paradigm shift in education?

*Questioning in this rapidly changing world
is a student's superpower!*

Preface

Why does asking more help students learn more? Studies done at the University of Davis showed that "curious subjects were better at learning more and retaining even unrelated information." According to Dr. Matthias Gruber, "Curiosity puts the brain in a state that allows it to learn and retain any kind of information, like a vortex that sucks in what you are motivated to learn, *and also everything around it.*" What happens in the brain when we ask a question? It searches through all the information it has taken in to find the answer, resulting in students accessing answers they weren't aware they knew. The saying, "Seek and ye shall find," is backed by neuroscientists' research like Dr. Gruber's. Prior to brain studies, I discovered that my students paid closer attention, **not** when I asked questions, but when **they** asked questions. When students are curious about what they are learning, the learning process is more pleasurable and they retain more of what they learn.

Why was questioning the game changer in my home and classroom? In the fifty plus years I spent in the classroom and the 47 years as a parent, I discovered the best way to serve my children was to empower them to use their questions to solve problems independently and to get their needs met from every teacher they encountered. They were allowed time to discover how they learned, and how to advocate for themselves, so their teachers could teach them *the way they learned.* These same skills have allowed them to thrive in their careers.

What did the parents of my students want? As most parents, they wanted the best for their children. They asked, *"How can I guarantee they will become successful, independent, and happy adults?" The 'question' was the answer?* They discovered *questions were* the *secret* that would lead to their children's success, so they shifted their role from homework helpers to homework coaches. They supported them as they learned to ask effective questions. The children thrived and their parents were thrilled.

Teachers asked, *"How can I meet the varied needs of all my students and create real engagement?"* The answer was to teach their children to use their questions to improve student interest. They quickly shifted their method of instruction to begin with students' questions. The results were their students experienced success in areas they never had in the past. As the students perfected their questioning, the teachers were able to cover more material and their students retained more with each question they asked. They were meeting their students' needs like never before and creating true engagement. This took the stress off of them.

Considering that innovation begins with questions, to empower our children to become the innovators of the future, we need to nurture curiosity by training students how and when to ask questions.

*"Children must be taught **how** to think, not **what** to think."*

—**Margaret Mead Ph.D,** an American Cultural Anthropologist whose work focused on problems of child rearing, personality, and culture

CHAPTER 1

Why is Questioning an Essential Skill in This Rapidly Changing World?

With the competition for attention created by technology, advertisers, social media, and politicians, it has become increasingly more difficult for students to become engaged and stay focused in class or when using their devices. Students' questions prove to be the most powerful tool to create desired engagement. When students seek an answer to a question, they listen.

Because COVID created huge gaps in learning, students need to be taught how to ask for support in the classroom. Asking questions for clarification will be essential to fill in the gaps resulting from the loss of the learning from 2020-2022. The stress the students experienced continues to have a residual impact on learners today, and punctuates the need for students to be asking questions to support their own learning.

Asking questions also proves to take the guesswork out of how to reach each child more effectively in the classroom. The questions students ask help teachers learn how each student operates, so the stress that was once experienced in learning new concepts can be easily remedied and instruction can become more effective.

Because of the nature of artificial intelligence, it's crucial our students hone their questioning skills. Artificial Intelligence can state true facts, but it often omits important details. For this reason, questioning effectively is imperative for 21st century students.

At no other time in history have our children been bombarded by information that appears to be, but may not be, 100% true. Learning to question everything they see on the internet will help them evaluate the strategies advertisers, politicians, or friends on social media use to grab their attention.

The world is changing at breakneck speed. An education focused on passing tests will not prepare students to deal with this change. For the last forty years, schools were evaluated based on a pupil's academic test scores. This forced many teachers to teach for test results and train students how to repeat information which left little time for critical thinking about information they were memorizing.

Starting in September of 2019, my students learned how to overcome their fear of asking questions and used questions to find solutions without the internet or parental help. By the time COVID hit, they had transformed into amazing problem solvers. This was the _key game changer in my class when we were thrown into COVID learning_. Their questions allowed my students to work smarter instead of harder. Because they had learned how to independently find solutions through trial and error, they were not afraid to tackle the COVID challenge. Their questions about how to manage their time in this in-home learning situation was easy for them to remedy. They had time to view a challenge and think about possible solutions that couldn't be found on the internet. Through trial and error, they finally found what worked for their unique needs and helped our class not only survive the sudden change, but thrive.

When a strategy personally benefits a student, they will use it. Students will quickly see how asking questions makes it much easier to learn concepts that once were very difficult for them. Homework also takes less time. Eventually students discover that once they ask a question, it leads to other questions and eventually guides them to independently finding the answers.

Students will maintain their attention until they get answers to their own questions. They discover the benefit of listening to others' questions, as well. Classmates often ask questions they wouldn't have considered. Other's questions make it easier to complete their own homework. _This results in a shift from passive to active listening in the classroom._

The question is essential in this day and age.

CHAPTER 2

Why Do Children Fear Asking?

According to Mark Victor and Crystal Hanson, in their book *Asking: The Bridge from Dreams to Your Destiny*, "We need to ask like children" to make our dreams a reality. The authors consider it a crucial skill for accessing our brilliance inside and attracting what we want. My experience in the classroom confirms this belief.

Unfortunately, children have developed a fear of asking questions. Next to dying and public speaking, asking questions in class is the #1 universal fear of students. FEARs are Feelings and Emotions that Appear Real but aren't. It is perceived to be real to the individual until we can show them another way to look at their fear and replace it with a positive mindset around asking.

To change mindsets about asking questions, we need to examine the causes for this reluctance. Since 1972, I have discovered several underlying reasons why students are afraid to ask questions in or out of class.

- *Early Childhood Experiences-Parental Influences*
- *Mixed Messages from Teachers*
- *College Student Reluctance*
- *Fear of Being Dropped from a Class*
- *Adopting Peer Bias*
- *Fear of Peer Ridicule*
- *Fear of Being Wrong*

Early Childhood Experiences — Parental Influences

Children develop fear related to questioning at a very young age. They are naturally inquisitive. Just ask a parent of a two-year-old. They ask an average of 350 questions a day, if they are not stopped by an irritated adult. Children ask questions to build understanding of what they are seeing, hearing, and about the world around them. When children are stopped from asking the 351st question, they learn their questions are a source of irritation and consequently, they stop asking. As a result, they forfeit the opportunity to perfect a natural skill necessary for achieving success and making sense of their world.

In this device-centered world, parents who focus on their devices when their children are trying to get their attention, send the message to the children that they are not as important as their parent's devices and their questions aren't important at all.

In her documentary, author Judy Blume shares how she knew at a young age that her mother did not want her to ask questions, because her mother would never answer them and showed irritation with her very deep questions.

A large number of parents and teachers feared asking questions in class themselves. Many were taught asking a question would make them appear dumb, so they never thought to encourage their children to ask. Many perpetuated the same belief.

One parent admitted, *"I never asked questions in class because I thought my job in school was not to understand everything, but to memorize everything I could so I could pass the tests. So, I never thought of having my children ask them. I'll be honest. I don't remember much of what I learned in high school and college, but I did get straight A's."*

Children are clever creatures. They learn through cause and effect. If they touch a hot stove, the pain teaches them to not touch. There are times when parents need to teach children how to avoid pain for their safety, but not all pain is bad for children. Forgetting one's homework, getting a bad grade because one plans poorly, or struggling with new concepts can all be painful at first. None of them are life threatening, yet parents try to ease their children's pain when they can. They may not be aware of how asking questions can relieve the pain of not knowing how to do assignments or projects. They don't realize how pain can lead to creative solutions. Because many parents provide answers or do projects, the children will not ask the necessary questions alerting teachers that their directions are not clear.

I recently heard a comedian do a whole set focused on how she was challenged by her son who came to her on a Thursday night at 8 pm to tell her he had a project due. When asked when it was due, her son replied, "Tomorrow." The audience of three thousand people burst out in nervous laughter. When she shared that the project was a 3D replica of the White House, the audience had obviously shared similar experiences because she had to pause to allow the laughter to calm.

Why would a student ask questions if they have parents who will reteach concepts, do their projects, or pay for tutors to do so? The message sent to the child is, "You don't need to listen in class or ask questions because I will help you." Consequently, children don't ask.

Mixed Messages from Teachers

One of the reasons students fear asking questions is due to how teachers have responded to their classmates' questions. One negative response by a teacher to one student's question will stop peers from asking. Students get mixed messages from teachers. I have heard teachers say, "There is no such thing as a dumb question," but the response indicated the opposite. Students cue into these responses and quickly learn not to ask in that class to avoid ridicule.

My Experience

I experienced a negative response to a question while taking a class with my very accelerated third graders. I asked a fellow teacher to come in and teach them Latin. Since I knew nothing about Latin, I was eager to learn. He encouraged the students to ask questions. One day I saw something that confused me. To model questioning for clarification, I raised my hand and asked, "I noticed that the sentences in Latin don't start with a capital letter unless they start with a proper noun. Am I correct?"

I thought it was a great question, until he let me know it wasn't. The teacher responded with, "How many of you can answer Ms. O's question?" All twenty-six 8-year-old students raised their hands. If I had been a student, that would have cured me from ever asking a question again. His question implied I wasn't listening or I would know the information like the other children. Even if he had been right, his response would have stopped me from ever asking again.

Because I had a hard time with the new sounds connected to letters, and I was concentrating on how he said a sentence, I missed the information related to my question. I had to model a good response to show the children how to overcome such a reaction by the teacher.

I asked, "I was focusing hard on the sounds the letters made and may not have heard what you said. Could you repeat the information for me please?" He validated my thinking. My students, however, were visibly disturbed by this series of events.

This proved to be a valuable teaching moment about how to respond to negative responses and was the topic of our next class discussion.

One student said, "Ms. O. I was so sad for you when he asked us if we knew. Did it make you feel bad?"

I confessed that even as an adult, I felt awful when all the third graders knew the answer but I didn't. I emphasized that the teacher really didn't know anything about me as a learner, so he wouldn't know that I had a hard time with the sounds connected to a new language. I knew if I told him why it looked like I wasn't listening, he would learn more about how I learn. I emphasized that whatever the response, I still needed to get an answer to my question.

Many teachers misunderstand what is going on in their students' minds as they are viewing instruction.

I reassured them that no one wants to be a bad example, and I was sure he had not deliberately tried to embarrass me, but I still needed to get clarification. It was up to me to advocate for myself.

Being in the student's shoes helped me keep my own responses in check.

Students need to continue asking until they fully understand the concept. I taught my students to ask for concepts in different ways. They embraced this and I learned multiple ways of teaching one concept. At one point I thought there were only 10 ways to teach long division, until Jena asked for more ways. Then I discovered there are 14 methods.

I used the same worksheets for over 50 years because it was perfect for eliciting questions, and every year I heard questions I had never heard before.

Some teachers may show outward signs of irritation that have nothing to do with the questions students are asking. Their non-verbal facial expressions may be misinterpreted as agitation because of the question, when the teacher's response may have nothing to do with what is going on in the classroom. It may be what the teacher is thinking about that happened on the way to school.

From a teacher's perspective, it can be frustrating when students ask the same questions that were just asked. Most students are not trained to listen to classmates' questions and only care about the question they are thinking about. Once they are trained to listen to other's questions, they need to learn to reframe the question if the answer provided does not clear up their confusion. They need to ask the same question in a different way to indicate they were listening. They need to request the teacher show another way of explaining the concept. This will result in forcing the teacher to offer another approach to teaching the concept.

Kevin's Mixed Message

Kevin was a bright, sensitive 8-year-old. He knew how important questions were in our class and became an insightful question asker. I was shocked when prior to a parent/student/teacher conference, his mother wanted to give me a heads up that Kevin thought I didn't like him. I was grateful she told me, because it gave me important information I needed to know about him.

When asked, "When did you get the impression, I didn't like you?" Kevin shared his perception, "I asked a question, and you had a funny look on your face."

I remembered the day he was talking about. I loved his question. It was thoughtful and demonstrated sophisticated thinking for an 8-year-old. I recalled telling him it was a great question. It was one of those days, when I did not feel well and should have stayed home. Any teacher will tell you that it is often easier to come to school with a headache than write up lesson plans for a substitute. By 2:30 pm that day, my head was pounding. When he asked the question, I was also thinking, "Why didn't you stay home today?" Apparently, he saw a different reaction to his question because of the expression on my face. I was also at an age when my resting face looked like I was unhappy. Even though I told him he asked a good question, he responded to my facial expression. I reassured him that I loved the question and explained how I was feeling that day. I asked him to not judge my response by the look on my face. It's an old face, and you can't take it personally. If you think I am unhappy with you it is better to ask, "Ms. O. are you upset about something?" instead of assuming you are the source of my irritation. This helped him in his interpersonal relationship with friends, as well.

He hid in the closet the last day of school because he didn't want to leave me. He is still a sensitive young man, but he doesn't take reactions personally and knows how to ask for clarification. After that incident, I made a concerted effort to smile more and let the class know when I was not feeling well.

That was a valuable lesson for both of us. Since then, I always told my students at the beginning of the year that they can't judge how I feel by the look on my face. This has helped them understand that a person's facial expressions may not be a true reflection of how they feel. They learned to ask for clarification from anyone no matter what facial expression they observe.

Teachers have no idea how they are coming across at times, so we need to train students to ask instead of assume.

College Student Reluctance

My program's goal was to train my own children and my 8 year old students to feel comfortable asking questions of their teachers, so they would be the ones visiting college professors' office hours. According to professors, it is a time for students to get free tutoring, but the most <u>unused</u> period of their day.

Older students with insecurities can be reluctant to ask questions, because they fear appearing like a 'kiss-up.' It is easier to start the art of asking questions as young as possible. It is difficult for older students to believe how powerful their questions are, because they have feared them for so long.

Once they begin asking questions, they begin to see how it leads to receiving more effective instruction and the development of rewarding relationships. These can translate into college admission letters of recommendation.

Joshua: A College Freshman

Joshua was a high school senior about to enter college. His mother was concerned that he was too dependent upon her. When I shared how the brain works, why he needs to ask questions for clarification, and how to use office hours to get his professors to teach to his style of learning, he was eager to learn how to ask. After my hour-long session with him, he said, "I wish I would have learned this in middle school. It would have made my life easier and less stressful." He emailed after his first semester and sang the praises of the simple trick of asking questions during class and office hours.

Fear of Being Dropped from a Class

A reason expressed by many college students for their reluctance to ask questions is the fear of being dropped from a class if they do. This is a mindset that thousands of graduate students share.

College Professor

Because of my experience as a teacher, I began encouraging my own daughters to speak with their kindergarten teachers privately when they were confused. The kindergarten teachers were so receptive, they never gave asking questions a second thought. As a result, they were comfortable visiting college professors' office hours. One professor shared how my daughter's questions changed his teaching. He thought his instruction had been clear for the 25 years he had been presenting the same lecture. This was until my daughter asked questions of him that indicated he hadn't been clear at all.

He realized he needed to reteach the concept in the next class session. "As I retaught the lesson, I could see the light bulbs of recognition in the students."

At the end of the session, he asked, "How many of you found this session cleared up any misunderstandings from the last lecture?"

Most students raised their hands. He asked how they were able to complete the assignments without this clarity?

Some students said they paid for private tutoring, while others relied on peer tutoring. Most of the students agreed that the way he taught the concept the second time was clearer and some of the tutors were incorrect.

It baffled him why only one student came to office hours for clarification. So, he asked them, "Why didn't you come to me when you were confused?"

One brave student gave him the answer, "We had to be in the top 1% of our graduating classes to get into this class. I thought if I asked questions, you wouldn't think I was qualified to be here." The professor affirmed that questions assured him he was being clear and that his office hours were there for private clarification. This was FREE private tutoring. He also clarified how one person's questions could make instruction clearer for everyone, so ask in class.

Daniel: A Grad Student

"I mustered up my courage to introduce myself to my professor after the first session in my graduate program. I felt stupid because I had never done it before. My friends wouldn't even try. But what they didn't know was that with every visit I made to the professor's office hours, the lectures that followed appeared to be designed to answer questions I might ask. My professor was learning how I processed his lectures. He was well-connected with corporations looking for promising students just out of grad school. Because of the relationship we formed, my name was the first he offered. I had several offers for jobs before I graduated. This was odd, since the job market in my field was saturated. To this day, I credit the time spent in office hours asking questions as the reason for these opportunities. There was no luck involved."

Adopting Peer Bias

Aria

One college student explained that he wouldn't ask questions of one professor based on the opinions of his friends who had the professor prior to him. "They said he hates questions."

After asking him to share more about the situation, it became clear that the professor was working under the assumption the students taking his class had just taken the prerequisite, so the information should have been fresh in their minds. The truth was most students had to wait up to three quarters before they could get into his class. Consequently, the prerequisite information was not fresh for them.

I encouraged the student to write the following email:

Dear Dr. S...

My name is Aria. I have tried to get into your class for the past three quarters. Your class is very popular and difficult to get into, so I am happy I am finally registered for next semester. I want to be prepared, so I need to know if I should review the book I have from the prerequisite or use another one?

Yours,

Aria....

Aria was certain, based on the impression formed after speaking with friends, that an email would anger the professor and get him started off on the wrong foot. To his surprise, the professor asked him to come to class the first day and introduce himself, so he could put a face with a name. Then he encouraged him to review the text from the prerequisite course. By writing the email, he was alerting the professor to the reason many students might have appeared to lack the knowledge he assumed they would possess. The professor thanked him for emailing him. This directly changed his view of the professor. Had he not contacted the professor, he would have perpetuated the myth shared by his peers.

The professor even stated the following in the first session of class, "Some of you may have taken the prerequisite some time ago, so I recommend you review it before the next class. What I will be sharing will be based on the foundational information from that course."

The rest of the semester he looked to Aria for validation that he was making sense. Aria made regular visits to office hours and developed a strong relationship with the professor. This led to a wonderful letter of recommendation to physical therapy graduate programs. He was able to beat out several hundred students vying for spots in graduate programs on his first try, while other students ended up waiting a few years to find placement.

Peer perception is not always accurate. Questioning other's perceptions will avoid forming incorrect opinions.

> *"If we change how we look at things, the things we look at change."*
>
> **—Wayne Dyer**

Fear of Peer Ridicule

A fear of being judged by their classmates is another reason students will avoid asking questions. Creating a learning environment that focuses on the unique needs of each student will help to eliminate this fear. We must remind students that we all learn in different ways, and they may need a different approach from another classmate. Sharing a variety of methods, allowing children to experiment until they find the ones that work for them, and nurturing the sharing of their success stories with their classmates will create a supportive rather than judgemental classroom. In such a classroom, it's okay to make mistakes, and it's expected that others might know what we don't know. It is never okay to ridicule others for not knowing. It doesn't make one student smarter than the rest because they know an answer.

People Pleaser Challenge

When students are people pleasers, they will be afraid to ask questions because they don't want to offend their teacher. If they do feel comfortable enough to ask, they might be reluctant to tell the teacher the answer given didn't clear up the confusion for them. Asking students if the answer cleared up confusion and reassuring them that there are other ways to explain the concepts if it didn't, will encourage these students to continue asking until the concept is clear.

Fear of Being Wrong

Perfectionists are afraid of offering a wrong answer or asking a question that might make them appear to be less capable. Some students would rather say nothing than reveal they are not as smart as people think they are. This is true for very bright children as well. *A teacher asked, "What do I do with a student, who is gifted, and who won't participate in class discussions? When I ask a question and give them a minute or two to think about it, he still won't participate."* A minute or two may not be enough for students who fear giving a wrong answer. They equate a wrong answer with not being smart, which puts them in a stressed state. They have a lack of understanding about how this fear can make it impossible to access any prior knowledge. It may take five or ten minutes for the stress to subside before they are able to access the answer. Many teachers think bright children 'should' be able to respond quickly. But the fear of being wrong will prevent even the brightest of students from being able to respond on demand. Even math facts can be impossible to recite if a student is in a stressed state.

When teachers help the perfectionist realize wrong answers or questions are an indicator of a need for clarification and not a reflection of intellect, they will relax and embrace the value of their questions.

CHAPTER 3

How Can We Help Children Overcome Their Fear of Asking?

Now that we understand why students might have developed a fear of asking, we need to examine how to unlock that fear, so they can become powerful question askers. In the previous chapter, suggestions were offered to help students recognize a different mindset they might hold. More information is often needed for children to overcome their fear as they begin to ask questions.

One of my editors shared that she was fearful of asking for a packet of ketchup at McDonald's. As an adult, she regrets that her parents didn't encourage her to face her fears, so she could overcome them. Parents will empower their children when they make them face their fears and ask questions for themselves. They quickly learn how to get their own needs met at an early age.

If Parents are fearful of asking questions, this fear can be passed to their children. They are less likely to encourage their own children to become self-advocates.

Much to the surprise of parents and students, teachers really cannot do mind reading, so they need the students to help them by asking questions.

When students understand how being curious about the subject they are learning will improve their retention, they may be more willing to ask questions.

Unfortunately, most students hold a belief that their peers will think they are dumb if they ask questions.

The following questioning process has been effective in helping parents and students help shift their mindset surrounding questions.

"Do you dislike asking questions?"

"Are you afraid to ask questions in class?"

Some children won't be willing to admit to their fear. It's important to point out that asking questions in class is one of the scariest things for most students, so they are in good company if they are afraid. I share that I used to be afraid to ask questions, too, until I saw how much easier learning became when I did. I reassure them they will not be judged for not knowing something. In fact, we will celebrate every question and answer.

"I understand that you might be afraid to ask questions, but I'm curious if you can think of a time when someone asked a question in class, and you were happy they did?"

Most students will remember one time when they were happy someone asked a question.

"Why were you happy they asked?"

The common response is that the classmate had the same question.

Only one time in all my years coaching students have I heard a 'no' response. At the end of this chapter, I will share how to address a 'no' answer.

"Did you think the person asking the question was dumb for asking?"

"No, I didn't think they were dumb."

"Why didn't you think they weren't dumb for asking?"

"Because I had the same question but was afraid to ask it."

"So, you didn't think they were dumb for asking the question, because it was the same one you had? Am I correct?"

"Yes."

"Okay, So, let me clear this up. Someone asked a question, and you were glad they did, because it was the same one you had, and you didn't think they were dumb.

So, my question to you is, why would you think they would think you are dumb if you asked a question that they might have?"

"Guess what. There are classmates who will be afraid to ask your

question, and they will think you are brave for doing so. You might ask a question they didn't even think about asking that will help them understand the concepts and make the completion of homework easier to finish. You are doing them a favor by asking. In all cases, my students have seen that the people who ask questions are heroes."

"Even if you are the only one who has a question, it still needs to be asked. There are reasons you might not know what others in the class do know."

With this new perspective, we will see students move toward a more positive mindset about asking questions.

When the Answer is "No"

I suggest we address the issue that is presented when the answer is "No" to the question, *"Were you glad your classmate asked the question?"*

"No."

"I am curious, why do you say that?"

"The teacher's response was cruel, I felt embarrassed for my friend. He said, 'If you had been listening to what I just said, you would know the answer to your own question.' I had the same question, so I wasn't going to say anything. I decided, 'I'll never ask this teacher a question.'"

"There are a lot of reasons why teachers might respond abruptly to questions. We'll be learning how to answer these kinds of responses from other teachers this year, but for now, you need to ask me. If you have the same question, I will teach you how to ask your question a little differently. So, from now on, please know that your questions are more important than anything else in this class. It's what I need to know to make sure I am clear to you. They are what will help you learn and retain more. If I look angry, please don't take my aging face personally. If you have a question, it needs to be asked!"

In the next two chapters, I share how to deal with negative responses and reasons students might not know what others know.

CHAPTER 4

How Can We Help Them Deal with Negative Responses?

Students need to have strategies for dealing with negative responses before they begin asking. If we are going to ask them to go into battle, we need to arm them with counters to negative responses.

Not all teachers understand how important questions are to improving their effectiveness in the classroom. There are teachers who simply want to lecture or read the dialogue provided in their teacher's manual, and they don't want to be interrupted. Some may be irritated with questions because it cuts into their instructional time. If this is the case, it's important for students to use the teacher's office hours to get clarification.

Students may hear an answer like my former student, **"If you were listening, you would know the answer."** An empowered response could be, *"I know it might look like I wasn't listening, but I was trying to understand what you had just said, and I got lost. Would you go over that last part again for me in a <u>different</u> way?"*

They may be told, **"I just answered that question."** A good reply would be, *"I heard the question, but I am still confused. Please explain step two. That's where I got lost."* Students need to be taught to rephrase the question and be specific about where they were lost or the exact part that caused them confusion. Saying, "I don't get number 2," will not be specific enough.

Or they might prepare the teacher for a question that might have previously been asked by another student. *"This may have already been asked, but I was concentrating, and I am still confused about what_____ means."*

Because teachers don't always know how students process information, an answer that clears up confusion for one student may not help another. Reframing the question with, *"I heard your answer and I am still confused. I remember what I can see and not much of what I hear, so could you draw a diagram to help me understand?"* When students understand how they learn, they will counter negatives more effectively.

In my book, *Create a Student's User Manual Using Secrets About Teachers Every Student Should Know*, students discover what they need to know about themselves, so they can know what to ask their teachers.

Teachers are people, too. They have stressors placed on them by administrators, parents, and personal issues outside of class. They get tired and, when not feeling well, can become irritable. The teachers may be frustrated because they think they were offering clear explanations when the questions demonstrate they weren't. Frankly, they may just be too tired to deal with another question. Just like I shared with Kevin, when a teacher shows frustration, students should consider privately meeting with the teacher to find out why the question appeared to be upsetting to the teacher.

Jonathon Wasn't Glad His Classmates Asked a Question

Jonathon was the pupil who was not glad his classmate asked a question and is much like most. His classmate was responded to rudely. I pointed out he could have supported his classmate, raised his hand, and let the teacher know he also had the same question. He wasn't comfortable with doing it in class but agreed to rephrasing the question and asking it during office hours. He reluctantly made an appointment with the teacher. He was actually scared to death to show up at the appointment. Jonathon mustered his courage and shared he also had the same question his classmate asked. We had rehearsed what he would say, and discussed every possible response he could receive with counters for each of them, so, he was prepared.. He was shocked when the teacher apologized for the response he had given and proceeded to answer the question. Rehearsing powerful ways to ask is important for dealing with challenging teachers.

My book, *Homework Solutions for Weary Students and Their Parents*, offers different ways to ask for support for a variety of challenges one might face in school.

CHAPTER 5

How Can We Help Reduce the Stress Related to Asking?

Even when armed with reasons why they should ask questions and ways to deal with negative responses, some students still fear asking. Offering reasons why students might not know something other classmates appear to know can ease the stress related to asking

1. They might have been on vacation or out sick when the concept was taught. Not being in class is a perfect reason for not knowing what other classmates do know.

2. They might not have been curious about what was going on in the class at the time this concept was taught.

3. The student might have been coming down with something at the time of instruction and it did not get stored in the long-term memory.

4. The students might have been concentrating on one thing said and missed information that followed. It's not possible to be focused 100% of the time.

5. They may have had a quarrel during recess with a classmate and were unable to concentrate once back in the classroom.

6. They could have been in the bathroom when it was covered. Teachers don't pause instruction every time a student goes to the restroom. So, there is a good chance students will miss a lot when they are out of the room.

7. The brain has a wonderful way of making space for new memories to be stored. If the brain doesn't think the information is

important to store, it will begin pruning the receptors known as dendrites to make more room for new information. After one of my lessons on how the brain forms memories, one of my third-grade boys raised his hand and said, *"I think my dendrites might have pruned around this. Can you go over it again for me?"* This helped him save face when he couldn't remember.

8. Some teachers are working under the assumption if a student knows the material, they will be able to give answers to verbal questions. Under pressure to perform, students can temporially forget prior knowledge. When students are called upon to answer a question they are not prepared to answer and aren't given time to think about, their bodies tighten up if this sends a message to the brain that a threat is near. The body goes into an automatic response that sends blood to the legs for a quick get-a-way. Until the brain receives blood to the areas that find information, the student will literally be brain dead. The fear of not knowing an answer will set off this stress response that drains the blood from the prefrontal cortex which makes it difficult to retrieve prior knowledge. When students are aware of what this response feels like and learn to relax, they can slow their heart rate and get blood flow to the area of the brain that will help find the answers. Asking the teacher for a moment to think about the question will help them find the answer they are seeking. If they become frustrated during instruction, it is possible that the stress would be the reason they don't remember concepts. They can say, *"I think I was stressed when we learned this and can't remember what you said. Please explain it again.""I need more time to think about the answer. Please come back to me."*

Second Grader

While observing a second grader while he was on-line during COVID, I overheard the teacher ask a question. The student raised his hand and the teacher called on him. She then asked an unexpected question, "What is your favorite movie?" This caught him by surprise, and he went completely blank. As soon as the teacher moved on to another student, the answer popped into his head. "I should have said, How to Train

Your Dragon." I shared what happened to him from a neuroscientist's perspective. He understood that the shock of being asked something he didn't have time to think about caused him to move into a stressed state. I suggested that next time that happens to say, "Ms._____ I need a minute to think about that, can you come back to me?" When I arrived the next day, he said, "Ms. O. The strategy worked. I wasn't ready for a question the teacher asked me. I asked her if I could have a minute to think about the question and she said, "Sure, just raise your hand when you are ready." He felt very empowered.*

Teachers may not be aware that students need time to think. They aren't aware that randomly calling on a student because they think the student isn't listening will throw the student into a stress response. The stress response will make them appear they weren't listening when, in fact, they were. The teacher will never discover that the student has the answer, until the student learns to ask for "think time."

Many students are auditory learners and won't look at anything that will cause a distraction while they are focusing on what is being said. This may lead the teacher to form the wrong impression of the student. Most likely, students who are called upon to provide answers when they haven't had time to think about an answer, will not appear to be listening.

They are educating the teacher when children learn to ask, "I know this answer, but didn't expect to be called on. Could I just get a minute to think and get blood back to my brain?" **Every question a student asks improves the teacher's effectiveness. Teachers will do things differently if confusion is brought to their attention.**

9. Another reason for not knowing what others appear to know is, at the time of the lecture, they might not have thought the information was important to store. In short, if the brain doesn't think the information is important for later use, it won't bother saving it in the conscious memory for later use. The information may be there, but the brain requires more time to access it. *One third-grader raised his hand and said, "I think this information is in my long-term memory and the dendrites have pruned, so it may*

take me a few minutes to answer your question."

10. Though teachers follow a curriculum for their grade level, not all of them cover the same material each year. Some years I got further in the curriculum than in prior years. Even though teachers collaborate and try to cover the same topics, they may deliver the material in a different way that sticks better with some students and not with all the students. The third-grade teachers only have a general idea of what the second-grade teachers covered. It is possible that a third-grade teacher will expect a certain amount of prior knowledge from students the second grade teachers never covered. It is likely only two of the three second grade teachers covered the material. When students come from different schools, they may not have the same background, and therefore, need to ask when they don't understand. Not knowing is not a sign of not being smart, it may just mean they were never introduced to the topic or don't remember ever learning it. They can share the following with their teachers, "I don't think the teacher covered this last year." If other students in the same class claim they were taught the topic, the student can save face by replying, *"My amygdala didn't let this information in last year."*

11. COVID presented a whole new challenge to students. Due to the limited time allotted for instruction and the challenges of transitioning to on-line learning, teachers were unable to cover aspects of the curriculum they were once able to in the past. Teachers didn't possess the skills to be thrown into such a learning environment. The nature of being on-line created attention issues for many children. Some teachers were more efficient with the new platform than others, so some classes may have understood more. It stands to reason there should be more questions now than ever before due to the gaps created by this disruption in learning. **Even two years later, there are still holes in understanding because of on-line learning, and the negative impact of isolation is surfacing. This is a not only perfect time but a crucial time to use their asking skills.**

Steven

Post COVID I worked with a boy who was in 6th grade. He was asked to write an essay on the first day of in-class sessions. When it was returned to him, it was riddled with corrections. It made my head spin just looking at the paper. He had no idea what most of the comments meant, because, prior to online learning, he wasn't in class when it was explained. I encouraged him to go back to the teacher and ask for one-on-one support. He, of course, was reluctant. I trained him to say, "Mrs. K, I had a hard time with Zoom learning, and I think I could use some help to learn what I missed this past year and a half. Could I meet with you during study hall to go over this paper? Many of your comments are not clear to me."

He helped her realize that she assumed he understood her comments, when in fact what she commented on was not covered during COVID by his on-line teacher.

Excuses will no longer be necessary to make students feel comfortable asking once students overcome their fears and discover that asking questions gets their needs met.

We've discovered how important questions are to learners, how to help them overcome their fear, and reasons they might not know what others appear to know. It's helpful to add just a little more value to asking questions from a neuroscientists' perspective.

CHAPTER 6

What is the Neuroscience Behind Asking?

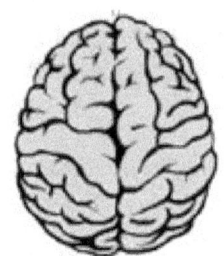

*"The important thing is to **not** stop questioning; curiosity has its **own** reason for existing."*

—**Albert Einstein**

Just telling my students they needed to ask questions or review material in short sessions over a semester was not enough. I knew there had to be another way to help them understand the importance of questions and the strategies I offered.

After attending a conference featuring the Dalai Lama, I realized the answer was to add neuroscience information to my curriculum. The Dalai Lama told a story about receiving an honorary degree at Columbia University. He was asked what he felt the university was doing right and where they could improve. He replied, "Teaching neuroscience and physics is what you are doing right. You also need to teach ethics. Neuroscience proves to the science minded what I have been trying to tell you for years. It proves everything I have been teaching."

To put it in simplest form, neuroscience offers a reason to ask questions. As stated earlier, brain studies have shown brain activity during question asking. According to Neil Cooper, neuroscience author, "When we're asked a question or when we ask one our whole brain is stimulated, and serotonin is released. The release of serotonin causes the brain to relax and makes it able to find answers and develop solutions. With the conditions set for the brain to respond to the question, there is a rush of dopamine. Dopamine plays a role in how we feel pleasure. It's a big part of our unique human ability to think and plan. It helps us strive, focus, and find things interesting. It's why we feel so good after asking a question and finding an answer."

When instruction begins with questions, we are allowing students to engage with material in a different way. The questions stimulate the serotonin, and the answers produce dopamine releases. Instruction becomes more interesting, satisfying and exciting.

The introduction of the basic neuroscience principles was the game changer for my students.

Neuroscience Explains Why Students Say, "My teacher didn't teach this."

Research by University of Notre Dame Psychology Professor, Gabriel Radvansky suggests that passing through doorways is the cause of memory loss. "Entering or exiting through a doorway serves as an 'event boundary' in the mind, which separates episodes of activity and files them away. Recalling the decision or activity that was made in a different room is difficult because it has been compartmentalized."

This explains why students can hear a lesson on a new concept and, by the time they get home, forget what was taught. Asking questions sends a message to the brain that whatever is discussed in class is important to save for later, so it won't be cast off as they walk out the classroom door.

Homework offers another reason not to discard the information. When students find the answers to questions during instruction, homework is less challenging. Discovering this makes them take the process seriously.

Sharing this information with parents calms their anxiety. It will also lead them to question statements such as, "The teacher didn't teach it."

Another time the brain decides what information to store is when we turn a page in a book, the brain summarizes what was on that page, stores important information such as highlighted words, and then discards anything that was unimportant. This makes room for the new information on the next page. It is more difficult to remember what is read online, because there isn't any clear beginning or ending of a page. The act of turning the page with one's hand sends a message to consolidate the information on that page.

Questions are a **doorway** to innovation. They signal the brain to seek any signs possible that will lead to an answer of a question. These answers can come in various and unexpected ways. The question, **"Daddy, why can't we see our pictures right away?"** lead to the invention of the Polaroid camera.

Reticular Activation System - RAS

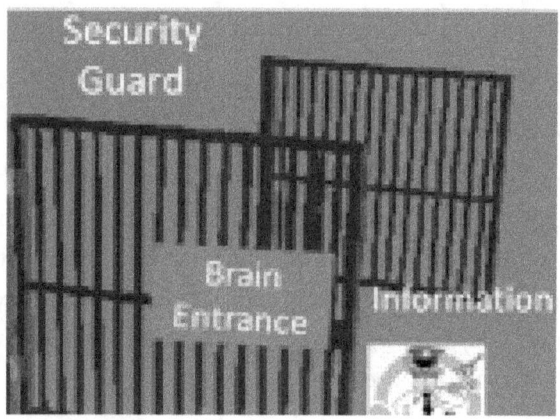

The brain's Reticular Activation System (RAS) is guided by questions to know where to send the information it's taking in. The RAS determines whether information is received at a conscious level or at a subconscious level. We absorb more information in our subconscious mind than we are aware of at the time. It is when we ask questions that we discover information we didn't know we knew. The questions we ask allow us to be conscious of what we might not have seen before. For example, I never noticed a Lexus SUV in silver, until I drove mine off the lot. Suddenly, all I saw were dozens of silver SUV's.

Researchers instinctively know questions provide the guidance system for the brain that leads them to make innovative discoveries. Their first step in researching is to list questions they hope to answer.

If our **RAS** didn't filter everything we see, hear, and feel, we would go crazy. Think about the engagement of a student who is watching a lesson and is confused. He forms a question for clarification. That student's **RAS** will be cued to listen for answers. Once students ask questions, they listen. And not only do they listen, but they do so actively with the intention of finding their answers.

Questions are the golden key to unlocking active participation and full engagement, and shifting the student from a passive to an active listener.

Neurons, Dendrites, & Hippocampus

Memories are transferred from one part of the brain to another through a system of neurons and dendrites attached to them. Concepts reviewed regularly have a strong neural path which speeds the transmission from the long-term memory to the working memory and back again. In the movie *Inside Out*, viewers are shown what happens at night when the brain doesn't get the message that the information is important. Cleaning men come through and throw the old unimportant memories into the dump to make room for new ones. Thus, students might do well on early tests, but by time the final is given, the memories are lost. Reviewing the information weekly for all prior units will prevent the information from being cast off.

Cramming for tests or being under the pressure to perform well on tests will create undo stress. This stress results in irregular formation of dendrites on the neurons that negatively affects the speed of transmission of information. Regular, stress-free practice will result in healthy dendrite growth and improve recall. Students who know this information will avoid cramming for tests.

Understanding the stress response, how it feels, and what happens in the brain under stress, gave my students reasons for practicing stress reduction strategies. They knew that calming their bodies while studying, doing homework, or taking a test will improve their performance.

Researchers at the University of California at Irvine studied the effective use of study time. They determined studying for one hour without breaks resulted in the subjects having one first and one last memory. These memories were easiest to recall. Further research determined that breaking an hour into four fifteen-minute sessions helped subjects easily retain four times more of what they studied. This bit of information promoted the nightly review of previously learned material. All resistance to such practice ended when they were equipped with the knowledge of how their brains functioned best.

Once the individual calms the stress response, their hippocampus structure is set into action to transfer requested information from their long-term memory or subconscious to their working memory. Students also saw the positive results of having a good attitude about finding answers while doing relaxation breathing when they were stressed. They often found answers they didn't realize they knew. *"The answer just came to me,"* is a common comment from students once they learn to tap into that infinite base of knowledge they can access when they relax.

Brain's Tendency to Confabulate

What do you see in this picture? If you said you saw two triangles, you would be ***incorrect.*** The definition of a triangle is a plane with three sides that form three vertices. There is only one triangle represented by this graphic. It's in the center of this image. If the white lines were black, then we would have one large triangle, but because they are white the larger image doesn't fit the definition of a triangle. The brain views the larger image as a triangle because it fills in the white space for us. This helps us make sense of the image. This demonstrates how the brain confabulates.

Confabulation is how the brain puts bits and pieces of information together to make sense of what it sees. What they see may appear to be true, though according to the rule, it isn't. Their brains subconsciously fill in the spaces. That is why people will believe they see a triangle when there isn't one.

Students can be certain they understand what was said in the class and not be aware that they are putting information together incorrectly. Those who ask teachers for clarification during instruction will avoid **confabulation.** Visiting office hours to make sure one is not confabulating will result in improved understanding and guarantee students will receive different instruction in the future. They are essentially teaching the professor how to effectively deliver material to them.

My Sisters

One way I can explain confabulation and how it impacts students in the classroom is to examine my own family. I have three sisters. Two of whom are close to my age. The third sister is considerably younger and therefore would not remember the events in our home as my other sisters and I do. My two sisters and I attended weekly visits to our grandmother's house for family dinners. We ate at the same dinner table with our uncles, aunts, and cousins. We heard the same stories being told at the dinner table. We rode in the same car, but to hear us retell the stories one would think we were at three different events. If you asked our cousins, they would have their own versions.

Our mom loved to sing songs while we drove from place to place. Even though we heard them and sang them at the same time, we have three different versions of the song, "While Standing in the Bathtub." Our mother's song was the original, and it was totally different from ours.

It is not hard to see how thirty-five students could be receiving the same information in 35 different ways. So, teaching them to ask questions for clarification is very important.

They think it is funny when I suggest they ask their teachers for support during office hours by asking, "I tend to confabulate. Would you look at my work and see if I am confabulating or if I'm on the right track?" I know the minute the student leaves, the teacher is going to research the meaning of the word.

When students understand how their brain works, they will employ strategies willingly and not think we are just giving them busy work.

Melody

A former third grader shared she was the only one going to bed at 8 PM the night before her college finals, while her roommates were cramming into the night. She outperformed all of them, because she learned at 8 years old how important it was to ask questions before, during, and after instruction. Melody also knew that reviewing written notecards weekly, beginning with the first chapter and adding subsequent chapters each week, helped her easily recall all the information by the end of the semester. Her neural path to the information in the first chapter and subsequent chapters was strengthened with each review. Melody also knew the value of visiting her professor during office hours, beginning after the first class, to be sure she was processing the lessons accurately. She still lived college life, but with less stress than her peers.

CHAPTER 7

What Method Can be Used to Promote Questions?

> *"The wise man doesn't give the right answers, he poses the right questions."*
>
> **—Levi Strauss**

Parents

Once parents understand how important questioning is to their children's cognitive development, they will be more likely to encourage their children to ask effective questions. They can begin by recording the questions children ask just like they record the milestones, such as the day they walked. Their questions will reveal much about the children, if we take the time to collect them and review them periodically.

By **not** offering answers to children's questions, we allow them to share their point of view before we share ours. We think they want an answer to their questions, but in giving our answer before we hear their thinking, we are limiting their view of the world to only our perspective. In many cases, they are just trying to put all the pieces together of what they are observing to make sense of their world. Carefully constructed responses can promote discussions that help them see there can be multiple answers to one question. Instead of answering the question, I encourage parents to pause and ask, *"What do you think?"* or, *"What makes you ask this question?"*

The solution to helping children form a healthy view of questions, is to encourage parents to take time to listen.

Encouraging children to think about their own questions and help them explore how to satiate their curiosity without the aid of the internet or anyone else, will increase creative thinking.

Second Grader

I was caring for a second grader who asked me, "Why are Tyrannosaurus Rex's front feet so short." I replied, "What do you think?" His answer was, "I think they are short because they are so big and they don't walk on them, but they need them to pull leaves off the taller trees. I think he is an herbivore." Honestly, I had no idea. I simply replied, "That's an interesting observation. I am not sure why, but you make a lot of sense. Let's research it and see what the paleontologists have to say." He was ready with a dozen questions every time I picked him up from school. His questions and reasoning demonstrated deep thinking that no one would have imagined. He was processing what he was seeing in his world. Had I answered his question, I would never have heard his ideas.

He had teachers who were frustrated by his constant questions, so I had to teach him how to ask in a way that would help them see his intellect instead of thinking he was annoying. I encouraged him to say, "I was thinking that Tyrannosaurus Rex's front feet were smaller than his back feet because............... Am I right?" Once he began asking his questions in this fashion, the teachers were impressed and opened up discussion with his classmates centered around his questions.

He is now in fifth grade and continues to ask more sophisticated questions about many different topics. Most center on how energy is created. His question this time was, "What is a better way to create electricity than solar panels or having to burn fossil fuel?" It's entirely possible that this question will lead to another innovative solution.

We need to help our children value their own thinking. Valuing their observations, even if they don't align with ours, builds confidence and results in an openness to others' perspectives. I have found it helpful to respond, *"That is an interesting observation, I see it in a different way. It's great we see things differently."* Using clarifying questions will help the child think more deeply about their own answers and serve as a guide for deeper thinking about future inquiries even before they ask others. If the answer is off track, clarifying questions are called for. A trip to the library to find other opinions can be a bonding experience for parent and child.

Recently, I had the opportunity to be asked questions about clouds by a boy who sat next to me on the plane. He asked, *"Will a cloud fall down if a plane hits it?" I responded with, "What do you think? He replied, "Yes." I asked, "Have you ever seen a cloud fall down?" I suggested he look out the window and watch what happened when we were flying. We didn't see any clouds falling down. After twenty minutes, he said, "I guess they don't. We just flew right through one and it stayed right there." I asked him, "What do you know about clouds?" He replied, "Rain comes from them?" "Do you know how the rain gets up there?" "No." I asked. "Have you ever seen the steam from a boiling pot of water?" "Yes." "Where do you think that steam goes?" "It goes up. Is that what a cloud is?" "What do you think?" He thought it did. He continued to ask a dozen more questions about how the water got there. We had an hour-long-discussion, every bit of which was driven by his questions. Through redirection, he discovered the water cycle. His mom said, "I really like the way you helped him answer his questions. I would have just told him about the water cycle. He really seemed more interested in asking more questions."*

We are in an era that is moving so fast many parents find it easier to give an answer, especially when dealing with frustrated children struggling with homework.

The parent, who smiles while helping their children, may achieve temporary peace in the home and may quiet the children for a moment, but it isn't teaching them how to independently find the answers.

Parents honestly think giving the answer is helpful, but they don't realize it creates dependence. It guarantees their children will need them for many more homework assignments to come.

We don't want to rob children of the true power of their questions. Rather than offering answers or instruction, parents can encourage their children to find clues in the chapter that could help them solve the problem.

Later in this chapter, I offer strategies parents and teachers can offer their children when they struggle.

Parents of my students requested support in making the shift from helicopter parenting to strategy coaches. This led to my first book, _Homework Solutions for Weary Parents_, now in its 2nd Edition, helps parents transition from Parent Teacher to Success Strategy Coach. _Homework Solutions: A Teacher's Guide - The Missing Link Between Home and School_ provides the same support for teachers. _Homework Solutions: A Parent's Guide - The Missing Link Between Home and School_ offers support to students and parents with over 40 different scenarios students might face during any given school year and how to ask for support.

Wondering Journal

Journaling is a wonderful and therapeutic way of processing our thoughts. It can also spark curiosity. It is an excellent place to record all the questions that pop-up for them. Who knows what they may lead to? J.K. Rowling recorded all her thoughts, and she ended up with many bestselling books and movies.

I encourage teachers to set the stage for questions on the first day of school. Asking students to write a list of things they wonder about or questions they have about the coming year will tell teachers a good deal about their students. I began by sharing my list of what I wondered about for the new year: Who will try something they never thought they could do and find out they can? Will I be able to help my students realize their talents? Will I manage all the administrative things I have to do and still be able to be the best teacher for them? What will be their favorite memory of our class? Will they be willing to make lots of mistakes so they can grow? Will I be able to overcome my struggle with writing report card comments? Will they continue to ask for support from their teachers when they leave our class? How many of my students will be willing to experiment until they find out their best study methods? How many of my students will help me by **not** asking their parents for help with the work I assign?

Examples of such wonderings are _"I wonder if I will ever remember my multiplication facts. I wonder if Jesse is going to bully me on the playground. I wonder if my teacher is going to get mad at me like all the other teachers?"_

Gloria wrote, *"Will Felicity still be friends with me? Will I have anyone to play with at recess? Will I finally correctly spell words? Will I be able to run the mile? I hate running!"* Darius wrote, *"I wonder why satellites don't fall out of the sky? I wonder where the water goes when it goes down the drain. I wonder what animal I will get for the research project? I wonder if my teacher is going to like me? I wonder if my mom will let me do my work by myself?"*

We learn a lot about how to help our students from their wonderings. I challenge them to record their wonderings about the things they see, hear, and read throughout the year. They are encouraged to record the answers they discover to all their questions.

In The Classroom

We need to provide students with a reason to ask questions and how to ask questions that benefit them. As was stated before and is worth stating again, one of the many myths' students have about teachers is that teachers are mind readers. Teachers don't know what their students don't know until the students let them know what they don't know. Teachers don't want to teach children what they already know because it will be boring. They want to teach them what they don't know. **It is the questions students ask that help teachers clarify where the confusion lies.** This leads to more effective delivery of information to each unique class of students.

If instruction is not clear and the students ask the parents, then the parents are the only ones who know what the students doesn't. The teacher will continue to use the same methods if the work indicates the students understood them. So, the teacher is the one who needs to know.

Because the question accomplishes so much, students need to develop the ability to perfect the art of asking empowering questions. They need to pay attention to the tone used when asking and how to be specific about what they do not know. It will become the most powerful tool available to them to learn more, remember more, and stay engaged. It will save everyone time by not having to reteach. Students who know how to ask questions grow into productive citizens of the world, while solving what may appear to be unsolvable problems.

Yet, we don't offer students time to think about things long enough to generate powerful questions or seek their answers. The desire to get through curriculum stifles questioning.

So, the first thing we need to do is allow time to learn how to ask. Initially, the process takes more time than teachers think they have, but once the students grasp the concept and master the art of questioning, they move through the curriculum much faster.

The following is a method I used in my classroom that helped students to understand the power of questions and how to ask them. I encourage parents and teachers to do the same. Rather than filling homework time with busy work, it can be used more productively as a time for question formation.

To engage my students, I asked if they would like their homework to take an hour and be really hard, or would they like it to take fifteen minutes and be easy. This provided motivation to listen. It has to benefit them to get buy-in. They chose easy. *"I don't want you to struggle so now is the time to look at your homework and find out what you don't know how to do."* They were directed to look at their homework before providing instruction, read the assignment **as if** they were going to **do it right away**, and circle the problems they could not answer right away. They were to highlight the key words that told their pencil what to do, and underline the words they didn't understand. I gave them time to record and ask questions about every aspect of the assignment. They thought they were cheating, but I knew the result of asking questions about what they were going to learn would create curiosity surrounding it, and result in them listening intently to the instruction. The dopamine release they experienced by asking questions and finding answers independently encouraged more questions.

I emphasized the importance of getting all the nuts about their homework before they left class by sharing my nuts-and-bolts story.

My Bookcase Adventure

*I went to the store to buy a bookcase. I traveled 15 minutes to get to the store, 15 minutes looking for the right bookcase that I could put together. Once I found it, I **casually** glanced at the directions in much the same way they looked at assignments and projects. It looked easy enough to put together. Then I waited in line for 15 minutes and drove 15 minutes back home. I arrived home and changed clothes. Then I prepared to* *put it together. At this point I had invested over an hour of my time. I proceeded to place all the pieces on the floor to put it together. It was then that I read directions carefully, "Place bolt part B into part A." I looked for bolts and found none. Then I read the rest of the directions carefully, (like I should have in the store) and in small print it said, "Nuts and bolts not included." I had to get dressed, go back to the store, find the correct nuts, bolts, and screws, and return home. I would have saved an hour of my time and one less trip to the store had I read the directions in the store **as if** I was going to put it together right away. What did I learn from this mistake? Next time, I will read all directions **as if,** so I will have everything I need to complete the task.*

This story helped them see how they could cut down the time they spent on homework if they made sure they had the nuts and bolts of the assignment by reading the homework questions *"as if"* they were going to do them **before I did the instruction**.

Once students understand the value of their questions, I ask, *"Whose question is the most important to you?"* They will respond, *"Mine."*

"You are partially correct. Your questions are very important to you, but the one's your classmates ask can be even more important when it comes to your homework. Why?"

After they have had a chance to share their reasons, they eventually come to the realization that others see things differently, and consequently, they will see something their peers might miss. So, they became very aware of the value of listening to their classmates' questions, and thinking about how they would answer them. They would ask themselves, *"What do I think is the answer to the question?"* Then they listen to see if they were spot on or if they need to ask the question in a different way to clarify their own thinking.

It's impossible to anticipate the questions one may have until they actually work on it. They discover what they should have asked when they start to tackle their homework.

Should challenges occur, parents can help them apply problem solving strategies to manage their homework instead of giving them the answers. The following questions will offer strategies they can use in the future, which will lead to independent problem solving:

1. Did you read the question aloud three times? (Reading the question aloud forces the reader to read each word in isolation and prevents an incorrect interpretation of the question).

2. Which words in the questions are unclear? Where can you find their meaning?

3. Can you visualize what the teacher wrote on the board? Can you see any clues? (One student shared that she took a mind trip back to the class and could see the lesson on the board. The part she was struggling with was covered by my head. She remembered exactly what I was wearing, but couldn't remember the last step).

4. Do you remember what the teacher said during instruction? Relax and try to hear what was said. (Some students have told me they can push their ear and replay the lesson word for word).

5. Are you feeling emotional because you forgot? Take a break and it will come to you.

6. Are you too tired to think right now? Go play for a little while and come back and try it again.

7. Try doing some deep breathing to calm yourself.

8. Skip the problem and come back to it. If you believe you will find it, the answer will come to you. It might come back to you after you have done the rest of the assignment.

9. A walk for fifteen minutes might help you come up with an answer.

10. Did you look for clues in the unit that might help explain the question?

11. Are there clues in the previous chapter?

12. Are there examples given in other chapters?

13. Did you try deductive reasoning by looking at the text for clues?

14. Did you use the process of elimination to cross out all answers that don't make sense?

15. Did you look at the homework like you were going to do it before the teacher began instruction, so you were familiar with the homework?

16. Did you ask questions during instruction to help you understand your homework?

17. Did anyone ask questions about this work in class? Did you listen to your classmates' questions and think about how you would answer them?

18. Parents need to avoid trying to teach the lesson. Instead, they need to insist that the children go in early to get answers to questions about what they didn't understand. Their teacher needs to know what they don't know. Ask them to write down questions they can ask the next day.

After each homework assignment, test, or project is completed and graded, parents need to ask their children, *"What did you learn from the mistakes you made and what will you do differently next time to avoid them?"*

As a parent, I found it helpful to tell my children I didn't know how to do any of their homework. It took them a while to determine that I did know, but by then, they had become the best question askers around and easily advocated for themselves.

Kim Saunders, a wise parent, and my Girl Scout co-leader, would say, "I can't fight your battle, but I will arm you so you can confidently go into battle."

Role playing how to ask questions when they went into battle was the key to success with my girls. To prepare my daughters for battle, I role played with my girls, I would shout out responses they coud use when a teacher might say, "Why are you wasting my time? I must grade papers!" or "Why weren't you listening?" They practiced saying, "I thought I knew how to do the work until I got home, but I realized it was harder than I thought. Can you explain how I do number one?" They were properly armed with a variety of responses. The teacher was never as mean as I was in my role playing. It was more important for them to overcome their fear of asking a teacher a question at age 5, than dealing with nightly homework challenges until they were in college.

I did alert the teachers the first time I had my daughters approach their teachers. The teachers were very receptive. When my daughters were in kindergarten, I called the school (we didn't have internet at the time) and spoke with the teachers. *"I want to give you a heads up. Shelly needs to talk to you about her homework, but she is afraid to do so. If she doesn't come up to you,*

could you ask her something that will open the door to asking her questions?"
That afternoon, Shelly skipped all the way home, came barreling into the house,
and announced how proud the teacher was of her to ask questions. "She even
told the class how proud she was of me for asking, and asked if anyone else had
a problem."

How To Model Effective Asking

Most students don't know how to ask empowering questions at first. They often begin questioning by saying, "I don't get number three." I would hold up three fingers and show them what three looked like. Identifying specifics about what they didn't understand is the first step. I had them dissect the questions to be sure they understood every word. "Do you understand what each word in the question means? After questioning each word in the questions such as, 'find,' 'the,' 'word' 'that,' 'is,' 'the,' 'for,' *'antonyms?'* They found the word that made the question difficult. They did not know the meaning of 'antonym.' I asked them to highlight the word or words in the rest of the problems that made the questions unclear.

My goal was for them to discover how to independently solve challenges like this.

A clever boy in my class raised his hand and said, "So what does it mean?"

I was glad he asked, but I wanted the students to discover how they could find the answer themselves.

I replied, "If you were faced with this problem tonight and you couldn't ask for help from a parent or sibling, you couldn't call a friend, you didn't have access to the Internet, how would you figure out what the word meant?"

I suggested they use the text to help themselves. This required me to provide enough time to review the clues in each question to determine what questions they needed to ask and find the clues that answered them.

I required students to come up with several possible answers for solving the problem and had them record them on sticky notes. Some quick thinkers raised their hands within milliseconds. I wanted to encourage them to think deeper about the question and see if they could come up with more ways to find the answer. It was important to allow all students time to think. Once every child had written something on their sticky notes, I reviewed all their answers. I asked the slowest responder to share first. This student normally would have stopped thinking because the others would answer quickly and there was no reason to give the question any thought. But the slower students' answers were often more creative than the rest.

Those whose hands shot up first and who were accustomed to being called upon were a little taken aback when they weren't the first called upon to answer. The class was surprised when the last student to record offered his observation, "I think antonym means opposite." He explained his reasoning. "I looked at our spelling word choices because it said, 'Find a spelling word for the antonym listed below,' I looked for words that meant the same, but none of the words did. But, there were words that meant the opposite of the words in the problem. Am I right?" I asked him if he thought he was right, and he was 100% convinced that he'd found the answer. His explanation made sense to his classmates, many of whom did not think of this strategy. Others shared their reasoning and the class learned different ways to tackle that same problem if it appeared in future lessons. One student used the dictionary for clues and didn't find it there and then looked in the Thesaurus and discovered the meaning there. I only validated his reasoning after everyone had an opportunity to chime in and share their solutions.

Using this method, students learned strategies from each other. I was surprised how many new strategies I gleaned from this method of instruction. The children are very creative when given the opportunity to think deeply about questions.

Months later, they were asked to find antonyms in a different program. They were all surprised how they remembered what it meant. If I had defined the word or they had looked it up in the dictionary, they would have been less likely to recall its meaning. The fact that they used deductive reasoning to discover the meaning helped them retain the information months later.

This method was time consuming in the initial stages of the program, but later allowed me to cover more material than in previous years. It also reduced their stress during test taking, because the strategies they learned during homework sessions were ones they could use when no one was able to help them.

This process was challenging at first for children accustomed to turning to parents for answers. It became even more frustrating once the students could get easy access to answers on the internet. This speaks to the importance of **not** allowing the use of any other sources other than their use of deductive reasoning, the process of elimination, and their textbook to solve challenges. Their book's dictionary had been unused until they had to become resourceful and use only what the book had to offer to solve their problems.

There are some teachers who believed I was giving the answers, but what I was doing was giving them a purpose to listen, question, and discover the answers themselves, and/or receive clarification. I never gave out the right answers, but guided my students to discover methods they could use at home to find answers independently. One parent described the results of this method as *life skills'* training.

The homework gave them a purpose to listen and personal feedback to them if their questioning was effective. They discovered if they needed to ask for more "nuts and bolts" the next day. It also provided me with feedback about where they needed more support to perfect the art of questioning. The beauty of creating questions to ask the next day is that the answer often appears before they put their homework away.

The side benefits of this approach were the parents no longer needed to participate in homework and the students were able to build a toolbox of strategies for later use, and perfect their questioning abilities.

I believe something <u>given</u> is <u>forgotten</u>, while something <u>discovered</u> becomes a part of us.

There are a variety of questions that stimulate the brain in different ways. To promote more than just clarifying questioning in the classroom, it's helpful to share the other types of questions. Allowing students to experiment with them by creating questions for chapter tests and using them in speeches requires critical thinking skills. Once they understand the different types of questions, deciding which ones to use for different purposes will require deeper and more complex thinking.

Using the story of *The Three Little Pigs*, students can practice writing different types of questions.

The least productive question that stops deeper thinking is the closed questions. It only requires a yes, no, or one word answer. This is the first type of question students will ask when they begin writing questions for their classmates. An example would be, *Did the pigs live together?*

The recall question calls for literal information. Primary students rely on recall questions. Story questions are simple such as, *Who are the main characters of this story? Where did the story take place?*

An open question requires the student to provide more details. *Why did the pigs leave their parents' home? What was the reason the pigs built different types of homes?* These questions call for motives and a simple yes or no will not suffice.

The process question leads to deeper thinking, stating an opinion, and examples to support one's answers. Reading the text with a process question in mind will help students focus on the material in a different way. Questions such as, *Why was it a poor idea to build a house of hay? Was it smarter to use twigs?*

Affective questions require an answer that shares feelings about a topic. *How did you feel when the first pig's house was destroyed? How did you feel when the wolf ended up in the boiling pot? Why did you feel that way?*

Problem solving questions ask for alternative ways to solve the challenges they are learning about. *What else could the pigs have done to protect themselves from the wolf?*

Evaluative questions stretch thinking by requiring readers to compare or contrast different opinions. *Support your reasons for deciding who was the wisest of the three pigs?*

The most frustrating questions for my new third graders were the inferential questions. One student stated, "Ms. O. I pride myself on always finding the right answer and there are no right answers to these questions." These questions required the student to draw conclusions using contextual clues. An example would be, *How do we know that the fireplace in the third pig's house was large? How do we know the water was not as hot as it could have been?*

Rhetorical questions appear in speeches and advertising and are designed to persuade. They don't require an answer. One of the activities my students did when studying advertising strategies was to create a new cereal to appeal to children their age and the parents purchasing the cereal. They were required to create a radio pitch to draw attention to it. They began each speech with a rhetorical question that asked about a unique feature of the cereal that would benefit the consumer. An example of one such question is, "Are you tired of cereals that get soggy in milk?" The question is designed to set up the listener to find out if there is another option for a cereal that doesn't get soggy. It isn't designed to get an answer. Once they write their own rhetorical questions, their reticular activation system alerts them to when advertisers use them.

Many children are fast readers. They may devour a novel in a few hours. Voracious readers often have difficulty recalling textbook material. Asking them to wonder after every sentence in their textbooks will improve retention. My students were reading about superheroes and how they were a model of good choices and morality. The topic sentence spurred the wonderings about how they could be a role model. When it stated the superheros were important during the Great Depression. The question they asked was, "What was the Great Depression?" Before they were asked to wonder about each sentence, they could not answer this question.

These readers often read the first sentence and it sparks a thought process that is not aligned with what they are reading. Their answers will reflect prior knowledge but not knowledge of the text. This method keeps their mind actively looking for the answers to the questions they form about the text as they read and prevents daydreaming. This process improves comprehension.

> *"The ability to ask questions is the greatest resource in learning the truth."*
>
> **—Carl Jung**

CHAPTER 8

What Strategies Can Be
Used to Find Answers?

"Learn from yesterday, live for today, hope for tomorrow. The important thing is not to stop questioning."

—Albert Einstein

Learning about the strategies Edison, Einstein, and other innovators used to unlock their brilliance and create innovative solutions is empowering for students.

Einstein said, "If I only had an hour to solve a problem and my life depended on it, I would spend the first 55 minutes thinking about the right question to ask. Once I knew the right question, I could solve the problem in 5 minutes."

Finding answers also involved silencing his brain in what is known today as a form of meditation. Einstein didn't turn to the internet for his answers, it didn't exist. He learned how to tap into his innate ability to find answers within. He didn't look to others for answers because he was looking for the one answer that hadn't been discovered.

I was sharing this information with a friend and her response was, "But Einstein was brilliant. That is why he could be so innovative."

She is correct, but his brilliance led him to realize that the *questions were the answer to finding his solution.* The character traits Edison and Einstein both shared that a growing number of students are lacking today are curiosity and tenacity. Edison did not give up after the first 98 failures. Each failure helped him clarify the questions he needed to ask. Edison kept seeking other solutions and each question led him closer to the final answer that resulted in us not having to read by candlelight or gas lamp. Creatives continue to utilize this method which has led to self-driving cars and drones. With the internet being a click away, students give up far too soon when seeking solutions. Immediate gratification is stifling their creative thinking. True creativity takes time, yet many students are asked to be creative on demand. Creativity doesn't work that way.

If we consider the number of failures Edison had, he must have asked thousands of questions before he found his solutions. Many students give up on asking far too soon. We want our children to develop the same tenacity and help them develop a plethora of questions they can ask. They need to ask until they understand.

Einstein knew that questions were the answer. He also knew that "To **FAIL** means **F**irst **A**ttempt **I**n **L**earning. **END i**s not the end, but **E**ffort **N**ever **D**ies, so if one gets **NO** as an answer, it really means **N**ext **O**pportunity." A. Einstein

Finding answers to questions requires that we have time to rest our brains so creative thinking can happen. According to Leonard Mlodinow and his book, *Elastic,* he explains how devices have robbed us of the ability to tap into elastic (creative) thinking. "This type of thinking results in innovations such as SpaceX or Tesla. It emanates from unfocused minds. A relaxed mind explores novel ideas; an occupied mind searches for the most familiar ideas, which are usually the least interesting." He points out, "The technological advancements that make elastic thinking ever more essential also makes it less likely that we'll engage in it. So, if we are to exercise the elastic thinking that is demanded by our fast paced times, we have to fight the constant intrusions and find islands of time during which we can unplug." Brains that are allowed to rest, function more efficiently. So, less homework is more efficient than too much.

Einstein learned to quiet his mind in a time that was much simpler and did not present the number of interruptions we experience daily. He rocked in a chair, while he held two metal balls, one in each hand. He placed two metal bowls below the arms of the chair. He would rock with questions in mind and relax his brain enough to lose muscle control. The balls would drop, and he would be awakened by the clinking sound with an answer or answers.

With the push to cover more material, we have over-burdened our children with hours of homework, which has left little time to rest the mind. Tired brains do not sleep well. Schools need to allow for time to process the material offered to their students and allow them time to ponder questions related to new information. Homework assignments can include strategies for accessing the creative mind while answering questions in a book. Homework is helpful when it is used to motivate students to ask questions. The work does not need to be more than 15-20 minutes per subject. Mindful use of homework will support children rather than cramming in curriculum that needs to be covered. Most of the crammed material is forgotten.

We need to take the advice of Leonardo DaVinci, "The greatest geniuses sometimes accomplish more when they work less." We need to help children tap into their ability by giving them time to stare out a window, take nature walks, or use the time to wonder about what they are learning. Some of my most creative ideas come when I am driving down the road.

CHAPTER 9

When and How Do I
Teach Children to Ask?

"Question Everything."

—A. Einstein

When Facing Adversity

No one is going to experience life totally free of adversity. Good questions can help one navigate challenges more effectively. One may ask, "Why am I always having bad things happen to me?" Or ask, "What can I do to learn from this adversity."

Bernie Marcus

Bernie Marcus, Co-founder, and owner of Home Depot was born shortly after his parents escaped persecution in Europe. His family arrived with the hopes of creating a better life for themselves which was impossible in their homeland. He didn't see poverty as an obstacle because of his parents' optimism. He recalls one of his first real obstacles. He wanted to become a doctor but the quota system in medical schools eliminated acceptance to college. He went to work for one company and the owner was his best teacher. Another obstacle was being fired as a CEO of a large corporation because he was a threat to the President of the company. This forced Bernie Marcus, a man who once lived in poverty, to create Home Depot. He looked at the mistakes his employer made and asked himself, "How can I do things differently and attract people smarter than me?" He dreamt of a mega store and eventually made Home Depot a success. He employed thousands of people who gave 110% to the company, because he made each one of them a shareholder. He realized it was important to share the wealth he was creating. Instead of being a victim, he did what his mother told him to do, "Move." Hundreds of stores later, and employees who became millionaires, one could say he learned from adversity by asking how he could do things differently.

Our students can learn to look at their adversity as an opportunity to make a positive impact because of it. They can ask questions like: What happened? What can I do differently to lead to something better? How can I make things better for others who are experiencing what I had to endure?

My youngest daughter had some health issues. Instead of complaining, she made it her passion to focus her research on what she thought might be the cause. Obstacles fell in her path that opened other opportunities she never dreamt about.

Using Artificial Intelligence -AI

AI is being used by students daily. They rely on AI to answer their questions, but they don't understand the nature of AI. They don't know how to determine the credibility of AI results because they have come to rely on device-generated information to be truthful and honest. Unfortunately, we are living in a time in history when anyone can post whatever they think and it suddenly becomes the truth, even when it hasn't been researched to determine its credibility. Artificial Intelligence has presented a new challenge for students and emphasizes the need to be a **good question asker**.

There are many views on AI. AI **isn't** designed to tell the truth according to Ezra Klein, a columnist. He shares there is a risk to our children related to AI companions, because the path to AI is smoother. Children develop relationships with AI that appear to be authentic but can become unhealthy. Ezra is concerned that AI is **not concerned** with the truth. It **cares about being convincing.** It doesn't matter that the information it provides isn't 100% truthful. "If the user cares about the truth, they need to be good editors."

AI may offer correct information but may leave out information that would change the results. The answers are in the details that are more often than not omitted. AI is only as good as those who input information into the program. It is up to the student to investigate who contributed to the information and if there are omissions and become good at determining any bias the creator might have about a subject. When they discover the omissions and bias, they quickly see how the information changes.

To help children understand how omission of information can impact our understanding, I challenge them to think about things related to the story "The Three Little Pigs." When provided with background information on the characters, the story changes. When one discovers that the wolf suffered from asthma and hay fever, it changes how we view his huffing and puffing. But this important information was never shared with the reader. What if the reader knew that he was vegan, or that the pig with the bricks was a bully and wouldn't share his bricks with his brothers. They were forced to use hay and twigs, which set off the asthma attack? This may seem silly, but AI is leaving out important information that would change the meaning if it was included.

The following summary of the 60 Minute interview with Google executives punctuates the importance of developing a list of questions to ask when using any AI or internet-generated information.

60 Minute Interview on the Topic of AI

According to David Byrne's 60 Minute interview: "In 2023, we learned that a machine taught itself how to speak to humans like a peer which is to say with creativity, truth, errors, and lies. Google CEO Sundar Pichai told us that AI will be as good or as evil as human nature allows."

"Bard, Google's AI, is here to help you brainstorm ideas, to generate content; a speech or a blog post, or an email," shared Google Vice President, Sissy Xiao.

Senior Vice President, James Manyka shared, "Unlike other AI, Bard does not look for answers on the internet like Google search does." David Byrne was astounded by the fact that, "Its answers come from a program that was virtually self-taught." Bard can process information 100 times faster than the human brain.

Bard appears to be thinking and making judgements, but according to James Manyika, Google specialist, AI can exhibit behaviors that look like there is someone behind it, but there isn't.

AI can create material, but as demonstrated in this interview, it is not always reliable. When asked to write an essay on economics, Bard fabricated the books it recommended. They didn't exist. This is a problem that the developers did not expect and has not been solved to date. When asked by Byrne, "How great a risk is the spread of disinformation?" The Google executive replied, "The scale of this problem is going to be much bigger." We learn that the problem with AI is it can create fake news and fake images easily by making it look like someone is saying something when they haven't. The viewers are cautioned that users must be responsible with its use. Some developers are trying to slow the release of aspects of AI, so they can get user feedback to build in safety layers. While others want a faster rollout.

In the short year and a half since this interview, AI Apps have multiplied, and lawsuits related to AI misuse are appearing daily. As I see it, users and viewers need to question everything, because they can't be sure if it is authentic or generated by an App. *The Neuron*, a publication focused on tracking new AI tools and researching about them, so the consumer doesn't have to, suggests that AI is creating more reasons not to trust it. Their research has shown that AI detectors' results are too often producing false negatives. In their publication on July 12, 2023, they stated, "It can't be trusted and furthermore, AI will soon outsmart any detectors. ChatGPT is already at that point."

Today, facts are offered out of context, consequently omitting important details. Interviews can be edited in a way that give the viewer the wrong impression of the interview. ***It must be up to everyone to seek the missing facts to get the whole story***. This requires one to ask questions. One might ask, what details did the author omit? Why did the author leave out so many important details? Is the interview presented in whole or has it been edited, which would change the tone and meaning? What was the author trying to prove? Did the author personally gain from omitting facts? Is the AI programmer biased about the information input into the program?

Many students, who turn to AI to write essays, need to be aware that the paper they turn in may be inaccurate. It may contain many correct bits of information, but it is omitting some facts that would change the outcome. All students using anything on the internet need to use many sources to determine the accuracy of the information they are using. My motto is, "Don't believe it, ask questions, and do your research!" There is so much we don't know yet about AI, so we need to be cautious. The negative side effects of video gaming didn't come to light until several thousands were addicted to them, so we are yet to see the full impact of AI.

Teachers need to question if the lesson plans created by AI take into consideration the unique needs of the students in the class. When we turn our creativity over to a program, we turn off that part of our brain. Will the lesson plans adapt to those teaching moments that happen that an APP can't predict? Will the APP know what type of students are being taught?

When Evaluating Standardized Tests

According to Dr. Ken Robinson in his book, *The Element,* Lewis Terman, the developer of the test to measure IQ, the Stanford-Binet test, "had an extreme view of human capacity." The test only measured one form of intellect. When viewing standardized tests, one needs to research and ask, "Who created it?" "How was it suggested to be used?" "What were the test creator's motives?" "Is this detrimental to the future of the test taker?" and "Does the number on this test measure one's other forms of intellect?" The Stanford-Binet is still being used in schools. I recommend parents and educators do their research on the creator's motives for developing the test. Colleges for years have used standardized tests to determine if students would be good candidates for their programs. It's a snapshot of one hour in a student's life. It does not measure the student's work ethic, which is more important to student success. It doesn't measure a student's curiosity which is directly related to later success. I would caution students not to place too much value on the score, because it doesn't measure every aspect of one's capabilities and creativity. Instead, I would recommend students ask, "What am I passionate about?" "What skills do I have that would make me good in the field of my choice?" "Is college really the right path for me?"

When Evaluating News & Commercials

Fake news has been around since the first newspapers and magazines were printed. It requires students to ask a lot of questions.

In the late 1850's, a magazine's readership was floundering. The publisher wanted to increase sales, so he conjured up a story about an astronomer who had a telescope so strong it could see Mars. The cover of the magazine depicted a tropical scene with fairies flying all around. Their scheme was successful, and they sold record numbers of magazines. The people believed the image was correct because it was in print. It has taken over 100 years to prove it was all a hoax. The magazine offered what appeared to be credible scientific proof. The readers never questioned who the scientist was or where they could go to see this planet for themselves.

I have been challenging my students to ask about everything they read, see, or hear since my first year of teaching in 1972. I found the best time to teach them about fake news was during election years. I brought in local and national newspapers. They were asked to watch the 5:00 nightly news. At the time, we didn't have news available 24 hours seven days a week. Our classroom discussions centered around inconsistencies they came across. There was some mudslinging on all sides. They could identify where articles were designed to get an emotional response. I shared magazines from prior elections and the students were able to determine that some of the politicians were opposed to some ideas in the prior election that they were promoting in the current election. They were bright sixth graders who quickly discovered that the politicians were like the advertisers we had studied. They could see the politicians create a problem and then promise to be the solution to that very problem. Being a critical evaluator of what appears on social media, internet platforms, magazines, newspapers, and radio is crucial in this world of 24-hour news. I recommend parents and teachers do the same today. The big challenge today is that interviews can be doctored to prove a commentator's opinion. Videos can be edited to mislead viewers. It is rampant today in all aspects of social media. My advice is do your homework. Review other interviews on the same topic and research the interviewee's history before believing anything controversial. Controversial usage indicates one side against another. Question all of it!

USA Today

An example of how the internet is a breeding ground for Fake News today is exemplified by a February 18, 2023, article in <u>USA Today</u>. According to the article, "A train derailment in East Palestine, Ohio, prompted an array of false and misleading claims on social media, where posts built on speculation, assumptions or outright falsehoods, accumulated thousands of shares and millions of views." The article proceeded to debunk five outrageous claims that unnecessarily created fear in the citizens in the area.

It's unfortunate that anyone can put something in print, and as soon as it hits the internet it is stated as a truth. In the case of East Palestine, not all citizens who read the postings also read the responsible fake news article posted by USA Today.

To reveal all the fake news on the internet would be a daunting task, so it is up to us to question what we read and hear, and to teach our children to do the same, so they don't live a life in fear of what they see on the internet.

Fake news doesn't only appear on the news, but it also appears on social media platforms all the time.

Emotional readers are not rational or logical readers. Students need to be trained to ask, *"Is this information designed to make me upset?"* They will be less likely to believe what is written on social media if they question if the motive was designed to get them emotional. If it does, they will be motivated to seek the truth.

When students are trained to question emotionally driven articles and commercials and know the questions to ask about them, they will make informed decisions about the credibility of information.

When Reading All Opinion Pieces

Opinion pieces state one person's opinion. Seeking other opinions on the topic is crucial. It is known that some researchers pick and choose data to support their position on a topic. Much like AI generated material, omitted data is misleading to the reader. For this reason, when readers see statements such as, "Some doctors,"or "Many scientists," one needs to ask, "What other opinions exist on this topic?" If textbooks don't offer a variety of opinions, it's up to the reader to seek them out.

The place to start when evaluating opinion pieces is to observe their own response to everything they read, see, and hear so they can recognize if the goal of the author is to create fear. Then they need to take calming breaths and ask questions about the author's intent and credibility. They need to seek each other's views. Unfortunately, not all views are available on the internet, so this may require a trip to a library to do independent research using other sources.

Educate Them in Advertising and Propaganda Techniques

The first rule of advertising is to identify what your clients need and want. They look for problems, the advertisers provide the service or product that will meet those needs or solve the problems.

The number one target of advertisers are children because they are the best influencers in a home. They appeal to children with their ads through the programs the children watch and the video games they play. These ads often appear in pop-ups. Children will become more critical of what they see if they are aware that the ads are attempts to manipulate them.

Jason

Jason had a terrible addiction to YouTube. He was fascinated by different aircraft. His mother reported she would catch him on his computer for an hour because the program would automatically start a new video when the first one was over. With each new video ads would pop up. I pointed out how those ads are there to get children to ask their parents to purchase the items. He said, "I don't click on them." He wasn't aware that the advertisers know that we don't have to click on them to influence us to purchase their item without us being aware. We are taking in the information at a subconscious level. The advertisers are aware of the neuroscience behind what they do. Reviewing the neuroscience around dopamine release is a powerful deterrent for young users.

Studying cereal boxes will also help children identify how advertisers influence them. The goal is to have children ask their parents to purchase their brand over others.

We were fortunate to have an advertiser who worked for a top brand-named product do a presentation for the third graders. When she began working for the company, sugar was a huge concern of consumers. So sugar was removed from the names of all their products. She revealed how the data from a study of a prestigious school had been manipulated to prove that sugar **wasn't** making us fat. Their scientific evidence claimed it was the fat content that was the source of weight gain. The companies added nonfat labels on the boxes. This information changed when it was discovered that we need fat in our diets. Over time heart disease, obesity, gluten, and immunity building labels were added as nutrition focus changed. Sharing this enlightens children, as young as eight years old, to critically view and question what they see and read on their cereal boxes. They become aware that data can be manipulated, and important data omitted making claims designed to sell a product seem more credible.

Testimonials from political figures and Olympians were featured to add credibility to their product, even though these individuals had no credentials for making any claims about the product. Major movie advertisers discovered the power of promoting their movies on boxes that targeted their desired demographic. Helping students identify and question the goals behind ads will help them to evolve into wise consumers.

The following are the questions students learned to ask about fake news, commercials, and anything they read, see, or hear on the internet:

Ads and Offers:

1. How will the advertiser benefit from this ad?

2. What does the fine print say about the offer or product?

3. Can I return this item free of charge? Is there a restocking fee or a return postage cost?

4. Are the characters on the box making eye contact with me? This is a deliberate strategy to build trust with children.

5. Where did they get their scientific facts? Are there other studies done on this product?

6. Does the brand that costs more taste any different than the less expensive, generic brand?

7. Have I seen this celebrity promote other products? What are his/her credentials?

8. Is the special offer benefitting me or the seller?

9. Who did the research resulting in the facts and data collected? "Nine out of ten people" would warrant questions like, Who were the people? How many people did they canvas? Was it a medical doctor or a different type of doctor? Was it an actor? What did the one person who didn't agree feel about the product?"

10. What buzzwords or strategies were evident in this product's ad?

11. Where can I find the study about this product's scientific proof? Are there other studies by different scientists?

12. Does the FREE ad require that the consumer provide an email or pay for shipping and handling?

News, Internet Articles or Videos:

1. Does this make sense to me?

2. What experience does the author have? Is their reporting accurate?

3. Is this author sharing something someone else said?

4. How does the author know this information?

5. Are there other articles that share a different opinion? For every opinion stated, there should be multiple opposing opinions on the same topic.

6. If the article says a whooping crane lays one egg a year and another article reports there are two, which one do I believe? Do I need to find different sources to validate the correct information?

7. If the article says, "Some scientists," are there other scientists who disagree?

Before Reading a Chapter or Receiving Instruction on New Concepts

The questions at the end of a chapter are indicators of the most important concepts studied. Reading them, before reading a chapter or hearing a lecture, primes the brain to know what to focus upon. Our mantra in class was, "Read the questions first, aloud, and slowly." Reading aloud also slows down how one reads questions and improves understanding, so the students answer what is asked and not what they think is being asked. Fast readers often misinterpret the question. If there aren't any questions provided, students are encouraged to create them based on captions, titles, and pictures provided. I was the pupil who would highlight everything in the book because I wasn't sure what was most important. When I discovered this method of preparing for reading, my performance improved, and my stress level reduced.

Math word problems become much easier to solve when students read them **aloud** three times. The first time is to find the question. The second read focuses on key words. Asking oneself, "Do I understand the key words and what operation they indicate to use?" Once we understand what we are looking for, the third read is to look for the parts of the problem that have the data we will need to solve the problem. By using the read aloud method, confabulation (or the misinterpretation of the question) is avoided. Fast readers often confabulate, especially if they read silently. This leads to unnecessary errors.

Many times, students read a chapter in a novel and can't recall anything because their brains went in a different direction even though they were reading the words. They need to deliberately think about what they are reading. Asking the following questions will promote deeper thinking and better comprehension.

After each paragraph they can ask:

1. Do I recall what I just read?

2. How can I summarize this paragraph?

3. What do I think will happen next?

4. How is this paragraph important?

5. What events occurred that present a problem?

6. Are there any words I skipped because they are unfamiliar?

7. Did I stop after each statement and wonder about what it meant?

At the end of each chapter:

1. What is this chapter mainly about?

2. What key events happened to build the plot?

3. Am I able to answer the questions I created before I began reading?

During Instruction

As I shared earlier, asking questions during instruction helps the students build firm understandings of the concepts. Asking for different ways of demonstrating the concept may be necessary. Having to explain concepts again and again can be frustrating for a teacher, so it's important to recognize that the teacher is trying to do the best job possible and needs more input from them. They can ask for support by saying, *"You are doing a good job teaching, but I am still confused. Is there a different way to explain this concept?" "Will you show me a diagram, picture, or model of what you are talking about, because I am still unclear?"*

I prefer my students ask me the questions when they don't understand, rather than asking parents when they get home. If a method appears successful, teachers will use it until they have students demonstrate it is not effective. Then they will try a different method.

If Forming Mental Images of What is Being Read is Difficult

One recall method I found helpful is to form mental pictures of what is written on the board and the vocabulary and drawings teachers offer. It can be used to access information in textbooks, as well. This allows students to take mental trips back to the classroom when they are stumped on homework. *I had one student share that she couldn't remember how to do the steps for long division, so she took a trip back to the classroom in her mind. The step she couldn't remember was covered up by my head. Apparently, she didn't* *realize I was blocking the step. We both learned she needed to **ask** me to move when she was making mental images of what was on the board, and I also needed to be aware that I might unknowingly block the steps from the students' view.*

One of the most important skills needed to be a good reader is the ability to form mental images of what is being read. Reading words and creating a movie of the actions described improves retention. To develop this skill, I recommend students use images in books and on the internet to develop the ability to visualize.

Some students think they can't visualize. To help them realize they do know how to visualize, they are asked to close their eyes and see themselves on a bike. Then they're asked to see themselves riding down their street on the bike. When they do this activity, they are ready to visualize what I write on the board and use the images to support them when they are doing their homework.

The following activities help students who need more concentrated training in how to visualize:

As they look at the images, they can ask the following:

1. What colors are present?

2. Are there more colors present after close examination?

3. Are there plants?

4. What shape are they? What colors are used?

5. Examine pictures of animals and magnify the image of their fur. In what direction do the lines go? What different colors do you see?

6. Notice the colors in the sky. Are there clouds, fog, or clear skies?

7. Is there a stream? What colors do you see? Are there rocks? What is on the side of the river?

8. Now close your eyes and see these scenes in your mind.

Some books describe locations and items that the reader is unfamiliar with and cannot visualize because they have never seen them before. Looking at images of these places and people will help make reading more enjoyable. If you have never seen a spaceship and the book doesn't describe it in a way that will help the reader understand what is going on in the story, the reader will need to do some research for images that support the author's writing.

When a Student Has Missed Instruction

When students go to the restroom, they need to ask questions about what they missed. When they know they will be out for a trip, sports events, or doctor's visits, they need to ask what they are going to miss and what they missed once they return. Having a study buddy is helpful because they can share notes when students are absent. Asking a few study buddies at the beginning of the year, "Will you take detailed notes about what goes on in class if I am ever absent or out of the room? I'll do the same thing for you." This will establish a relationship that one can depend upon.

To Improve Performance on Papers and Projects

As mentioned earlier, I encouraged my children to use office hours. My daughter had taken an AP psychology class in high school. She received an "A" on a research paper. In her freshman year of college, she felt it would be best to take the equivalent class to make sure she had a firm foundation. The professor assigned a paper on the same topic as her high school paper. She knew to ask for feedback on her paper prior to the due date. She gave herself a decent window of time for the professor to review her paper and still have time to rewrite, add needed information, or find more resources.

I received an upsetting call from her when the professor returned her paper. It was riddled with red marks. Her first reaction was that it indicated she didn't belong in that college. What she discovered was, the professor handed her back what he would consider an 'A' paper in his class. His standards were different from the high school teacher's standards. He was looking for something different. She made the edits he suggested and learned a lot from the red marks.

I always felt that having students turn in papers to see if they were on the right track got them to look at the corrections. If they turned in a paper without asking for this support, they most likely would have seen a "C" or

"D" grade and thrown it in the trash, while not looking at all the comments the instructor made. That is a true waste of an instructor's time. Getting the "C" paper while they are still writing it and before it is due in its final version, promotes learning from mistakes.

Questions to Ask Oneself About Errors on Homework and Tests

1. What mistake did I make, and how can I avoid it next time?

2. Do I need to ask more questions next time?

3. What questions could I have asked before I began my homework that would have helped me avoid this mistake?

4. Did I rush to do my work?

5. Was I too tired when I did this work?

6. Did I pay attention to my classmates' questions?

7. Did I reread the questions aloud and double check that my answers answered the questions asked?

8. Ask the teacher the next day, "I am not sure what this word means on this test. Will you explain the directions because they are unclear to me?"

9. Ask the teacher, "I double checked my work and I think I got the right answer. Will you look at it and tell me where I went wrong?"

Daniel

A third grader had a problem marked wrong on his worksheet. He said, "You marked this wrong." Approaching teachers in this manner will put some on the defensive. I encouraged him to reword his statement, so he could get the teacher to realize why the problem was marked incorrectly. He rehearsed, "I was going over my worksheet and I couldn't figure out why it was marked wrong. I tried the problem three times and kept getting the same answer. Will you please show me where I am going wrong?" This approach will help the teacher recognize a grading mistake or see what further instruction the student needs to have a firm foundation for concepts that follow.

Michael

I warned the students to pay attention to key words in questions. These are words that tell their pencil what to do such as circle, underline, put in alphabetical order, or cross out. They were also alerted to focus on the key adjectives in the questions, such as a <u>long</u> *vowel sound,* <u>short</u> *vowel sound,* <u>smallest,</u> <u>largest,</u> *etc.'*

After reviewing his social studies test, Michael, one of my third graders, approached me and said, "Ms. O. I think there might be a mistake with the grading key, because I double checked my answers, and you suggested I was missing a canoe as one of the items the Kwakiutl made from cedar. The question asked for small things made of cedar and a canoe is a large thing."

He was correct. I discovered I was using an incorrect answer key. Teachers don't double check their answer keys for correctness, until a student inquires about a question they think is marked incorrectly. Michael alerted me to the fact that the answer key was incorrect, and others may have had answers marked incorrectly, when they weren't. I made a point of telling my students there will be times when I accidentally or deliberately mark something wrong and it's up to them to check their answers and advocate for their choice. They were told they needed to explain why they thought the answer should be changed.

Test results offer students clues about what they really understood and what was unclear. After reviewing graded work, students need to ask for clarification. Understanding where they confabulated will help strengthen understanding and build a stronger memory around the errors.

Identifying one's mistakes and recording what was learned from them, and planning ways to avoid the same mistake in the future solidifies understanding and avoids making the same mistake again.

My homeschooling student made one of the most common mistakes made in math problems. She added instead of subtracted. After she saw her mistakes, she failed to record what she learned from them. Consequently, she made the same mistake the next day. When asked what she learned from this mistake, she stated, "I didn't look at the operation I was supposed to do." She was asked to record the mistake on her paper using a green pen and prescribe what she will do next time to avoid making the same mistake. She wrote, "I didn't look at the sign before doing the problem. Next time I will circle the operation and say it out loud." She didn't make the same mistake again.

When Assigned a Project

As soon as they receive a project, students need to look at projects 'as if' they will begin it right away. This process can be used once they enter their careers.

Questions they need to ask are:

- When is this project due?"
- Do I understand every step of the directions?
- What items need more instruction?
- Do I know how to do the activities I am asked to do?
- What supplies am I going to need?
- What help do I need from my parents? (The only help they should need is transportation to get supplies).
- How much time will each of the steps take to complete?
- What challenges have I had in the past that I need to avoid this time?
- Is there enough time to complete this project by the due date?
- What outside activities will get in the way of me completing this assignment?
- What tests or reports do I have coming up that I must also plan for while I am doing this project?"
- Do I have enough ink in my printer to complete any typed work?
- Are there any athletic activities that will take up a lot of my time?
- Do I have enough paper to print the report?
- Do my.parents have trips planned that will take up time?
- Do I have other projects for other classes?

I address this in detail in my book, _Backwards Planning: A Cure for Procrastination_.

During the report process, students need to touch base with instructors. It is best to do so by giving the teacher plenty of time to read it and give feedback and still be on time. They can ask the following: "I want to make sure I am on the right track with this project. Is there a good time for you to look at it and tell me if I am on the right track?" At that time, the teachers will tell them if they are missing concepts or if they need to rework a portion. Or better yet, tell them they are on the right track.

Johnny

Recently, I was working with a student on his writing skills. Johnny was not fond of writing nor reading. We reviewed a graded writing sample. There were many comments made, a few of which were illegible. There was also a rubric typed on red paper that was very difficult to read. He admitted he never looked at the comments because he was only concerned about the grade.

This is normal for students. Unless their grade is going to improve by making corrections, most will never look at the comments.

There were 12 comments he did not understand. We had to reconstruct the rubric to make it easier for him to process. It took an hour for him to form clarifying questions for each of the comments. He realized that the rubric would be easier for him to read if it were printed on different paper with fewer words, so we cut the rubric apart and he summarized the comments in a way he could later use to review his work before he turned it in. He was encouraged to add to the list of questions, "Could you print this on white or beige paper for me? I have a hard time reading it on red paper." He realized if he asked the teacher to review his essay a day or two before it was due, it would improve his writing. Then he'd be motivated to read all the comments and make corrections before it was due. In the future, he will read the comments on previous essays and ask questions before he continues with new assignments.

During Discussions

"Asking a good question can be valuable in and of itself, irrespective of the answer. It communicates your respect for the other person." Adapted from the Iowa Peace Institute Message

The following are examples of questions one can consider asking to clarify statements others make during discussion.

1. I'm curious, what clues from the story or chapter helped you think of your answer?

2. Am I correct you said, "_____" (Paraphrase what was said)

3. I understand your statement that _____. Is it possible the clues could also mean _____? (provide a different perspective).

4. What made you think this was the answer?

5. I am not sure what you mean, will you explain it in a different way?

6. I want to make sure I understand what you said. Is it _____?

7. You referred to _____ when you shared. Will you please explain a little more what you mean by that?

8. When you said _____, did you mean _____?

9. Will you please repeat what you said? I want to be sure I heard you correctly.

10. I think you said _____ Am I correct or am I off base?

11. Could you be more specific? Your question could mean so many different things.

When Attending Higher Education Classes

There is a lot of value in attending college if it is something one needs for a career they are interested in pursuing. It offers students supported independence and an exposure to many new ideas. However, students need to be able to think for themselves when they hear lectures that are not in line with their family values.

When I began teaching, we were in a year of a significant election. I can remember our principal, Al Sanchez, clearly stating, **"As educators, we may have our own views on who should be elected. It is not our place to share our personal views on any topic, but to allow our students to explore all of them.** If asked what your opinion is, let your students know that your job is to help them explore all the opinions. Yours isn't for sharing. Discussions can be had sharing a variety of views, but **at no time should I ever have a parent in my office complaining about hearing their child share yours."**

My First Teaching Experience

I was excited to begin my first teaching experience as a full-fledged credentialed teacher. I was armed and ready because I had memorized all the phrases recommended by the prominent child psychologist at the time. After two minutes in the classroom, I had this horrible epiphany! The students hadn't read his book, and they didn't know how to respond to these so-called 'magic phrases.' I drew the conclusion that the professor hadn't tried his theories on children. I almost turned in my resignation because I realized I wasn't prepared at all. After those two minutes, I had to rely on my own resources to gain my students' respect and get them to be respectful.

In today's colleges, many instructors share personal views on topics. Many of their views can be on controversial topics. What we want our students to know is, even though an educator has a view, it isn't necessarily for everyone. We need to remind our students they should expect to hear a variety of views and the instructor's opinion is just one of them. If different views are not offered, one needs to question that. Some educators' views might be different than others, it doesn't make them right, but what they are offering is simply a different view.

When my daughter was in her residency to become a Dr. of Physical Therapy, she used her skill of asking for clarification to question the recommended protocol for the patient they were studying. *"I understand your recommendation. I am wondering if....would also be appropriate protocol for this patient?"* When the doctor left the room, her colleagues asked her how she had the courage to ask that question. She replied, *"I guess you could say I learned it in kindergarten."*

Questions to ask the instructor or oneself:

1. What other views are there on this topic?
2. Why is the instructor not sharing a variety of views?
3. Are there other theories I need to investigate?
4. Do I see any gray areas in the issue being discussed? Ask about what is unclear.
5. What facts, if proven to be true, might cause us to think differently?
6. What studies led the instructor to believe as he/she does?
7. Ask, "I researched this topic and Dr ____stated _____which is different. Can you speak about this difference?"
8. Is this instructor's position in line with my religious beliefs?
9. Is this instructor's position in line with my political beliefs?
10. Is this instructor trying to change my core beliefs?

When Someone is Being Mean

Many children are faced by peers who are mean to them. Children, including young and grown adults, tend to take everything personally, until they realize a mean person is sad, not feeling well, or jealous about something.

Questions can diffuse tension, if someone says something unkind to them, they can ask:

1. "Are you upset or sad about something?"
2. "Did you mean to hurt my feelings by what you just said?"

When a Student Has a Learning Challenge

Many children have learning challenges. Instead of making an excuse, we need to help children discover ways to overcome their challenges.

After extensive testing, many students end up with an Individual Education Plan or IEP that provides direction to parents and teachers to improve the student's educational experience. Unfortunately, this is rarely shared with the student. I felt demystification was very important for my students. Understanding why something is hard can help improve a student's self-image. Equipped with strategies and ways to ask for support for what they need, their challenges are diminished.

We can often discover a lot about the way children learn in our classrooms by asking them a few questions. We will help them find out where their strengths are so they can use them to support their own weaknesses. Instinctively, students can answer this question, "Do you remember more of what you <u>hear</u> or <u>see</u>?"

If students are not auditory learners, they need to have visual support in the form of illustrations, diagrams, or demonstrations of math processes. Most of what is said is lost, so these students need to focus on what is written on the board. If nothing is offered as support, then the students need to ask for it. Learning to ask, *"I am a visual learner. Would you please show me what you are talking about? Is there a picture that will explain it?"*

If children are auditory learners, they may avoid looking at the board while processing what is being said. This can lead teachers to believe they are not paying attention, when in fact, they are. When students understand this about themselves, they can prepare the teacher at the beginning of the year by sharing, *"I am an auditory learner, so I might not look at the board when you are talking. Is there a way I can copy my classmates notes or you could capture a picture of the board, so I can review it? This way I can focus on hearing what you are saying."*

One 5th grade teacher told me, "I know exactly who was in your class in third grade. They come to me the first day of school and share:

- *I need to sit in the front row so I can concentrate and prevents me from talking to my friends.*

- *I'm an extrovert and will do better if I am away from my friends, but I am seated next to my best friend. This might get me in trouble.*

- *I have a very busy schedule after school. Could you let us know what* the homework will be before recess, so I can ask questions about the assignment and get started on it during recess and lunch?

- I get stuck sometimes when trying to write a story or a paragraph. I found a little walk to the chain link fence helps me gather my thoughts. Could I do that in this class?

- I've had dozens more ask for different kinds of support. The one thing they all do is ask for clarification a lot."

I couldn't have been happier to know they took me seriously when I told them they were in control of their own education, and they couldn't rely on teachers reading their minds.

Maddie

Maddie was identified with a form of dyslexia at a young age. Her strength was in her auditory skills. To support herself, every year, she approached her teachers and shared her challenge. She let them know she wasn't asking for special accommodations or extra time. Maddie learned at a very young age to use office hours to tell her teachers what she needed. She didn't make herself a victim, but instead, she became a powerful self-advocate. She continued to self-advocate through high school and into college. If she was not sure about a concept, Maddie would ask for clarification during office hours. She recently shared that she has found that her teachers want to help and when she told them what she needed, they were eager to provide the support. In one such case, she asked if one of her professors could post his PowerPoint slides on the internet, so she could review them after class, since she was more of an auditory learner than a visual one.

Chris

Chris was a sophomore in high school. He was a good Spanish student, so he took Spanish III. For the first time since taking Spanish, he was failing. The teacher taught by totally submerging the students into the language by only speaking Spanish. He did not know how to make *sense of what she was saying, because he was not an auditory learner. As a visual learner, he needed visual support to attach to the words he was hearing. The teacher felt submerging the students was the best way for them to learn, but for him it was resulting in one failed test after another. She didn't know how to support him. It was up to him to advocate for himself. We rehearsed what he would say. "Senora, I am not an auditory learner, and I am lost because I don't understand the words you are saying. I need to see what the words look like so I can understand what you are saying. Could you write the list of words you will be using on the board or give me a list of words I can use for reference?" She agreed and discovered he was not the only one who benefitted from this simple change.*

> *"Everything we hear is not a fact. Everything we see is a perspective not the truth."*
>
> **—Marcus Aurelius**

CHAPTER 10

How do They Find Answers?

Start With a Positive Mindset

O nce they've overcome the fear of asking a question, they need to allow the answer to appear. Asking us to stop thinking is like asking our heart to stop beating. To find answers means we must settle our brains and allow our brains to focus on finding the answers. Our intentions need to be deliberate. "Intend to Attend" was a motto in our class. We need to **believe** the answer will appear to us. This requires us to deliberately block our distractions. A form of meditation allows for our distractive thoughts to float away like clouds allowing our creative minds to find answers. The answers will appear in ways one would not expect, so we need to stay alert for them.

The Importance of Celebrating an Idle Brain to Find Answers

Dr. Nancy Andreasen, a neuroscientist, rediscovered the peculiar patterns of electrical energy by the idling brain. Her research proved that the mind that seemed to be at rest, was in fact, very busy.

Leonard Mlodinow describes the incredible activity that goes on between many structures in the brain when it appears to be at rest. "Reverting to the inner selves, the neural networks of our elastic thoughts can rummage around the huge database of knowledge, memories, and feelings that are stored in the brain, *combining concepts that we normally would not put together,* and noting connections that answer our questions. That's why resting, daydreaming, and other quiet activities such as taking a walk can be powerful ways to generate ideas."

The students quickly learned that if they were stumped, they could take a break, do some cleansing breathing, or wait until the answer came to them. Some found answers after just a short run to the end of the playground. They discovered that often a great time for them to ask questions in solving big problems was right before bed. They would come into the classroom very excited the next day when one of their answers came when they first awoke or when they were showering. I often ask myself questions before I go to sleep, and the answers often appear in a dream or upon awakening.

The statement, "Seek and ye shall find," is often heard as a spiritual statement. Neuroscience has proven that the brain won't rest until it finds answers. Many people give up too soon. Ask a question, quiet the mind, and the answers will come. As students form questions to ask the teacher about what they don't understand, they often find the answers without help.

Quieting our minds like Einstein did is very difficult today with all the noise around us. We may be trying to solve a problem, we start to quiet our minds, and the cell phone rings with a text message or a phone call. We can't access our answers if we are thinking about a dozen things.

If we try too hard, we make it more difficult to find the answers. It's hard for our brain to stop thinking. It might be pointing out negatives that draw our attention in a different direction. We want to balance our thinking with pros and cons and tap into our intuition. When I stop thinking too hard about finding an answer and do something like gardening, the answer often pops into my mind. If we don't anticipate answers, we can tap into what the answer can be— and it may be different than the one we expected.

According to Jeffrey Allen, author of _Duality_, "The analytical mind sees details and precision. It assembles information, and a plan about how to get from one place to another. It also is necessary for creating software. It's the part of the brain that puts parts together to make sense of things. But it can get in the way of creative problem solving."

Allen states, "When the analytical mind is quieted, it allows for the brain to see the Big Picture, get a Holistic understanding, a knowing of which way to go, that appears wise. The analytical mind doesn't allow us to dream. Creative thinking is like dreaming without boundaries. Our thinking mind needs to be quieted because it stops us from believing that we can find answers. We must quiet it, which allows us to find the answers if we believe we can."

Some students find it difficult to tap into their innate problem-solving abilities, because they overthink the answer. They can learn to tone down their analytical thinking by listening to music, or doing something physical, thus allowing themselves to simply quiet their mind. "Once they are calm, they can solve their problems. The calming allows them to see the forest for the trees. When they are being analytical, they just see the trees," states Jeffrey Allen, intuition specialist.

CHAPTER 11

How Can Teachers Create a Questioning Environment?

I love Q & A's and circle discussions. To promote and show that we value questions, it's imperative that clear standards for group meetings are established.

I asked the class to work in small groups to collaborate and come up with a set standard to keep in mind whenever anyone asks a question.

The following are ones they came up with when someone asks a question:

1. Respect each person's question, even if you think you know the answer.

2. Listen to each person's question, because it might be something you didn't think of asking.

3. Only positive comments can be made regarding the questions.

4. You can ask for clarification about a person's question, by reframing the question. For instance, "I think you are asking this, am I correct?"

5. Think about how you would answer the question to see if your thinking is correct.

6. Thank classmates for asking the question, since it will help us check our thinking.

The following is the list they developed during classroom discussions to ask for clarification:

- Make eye contact with the speaker.

- The speaker needs to make eye contact with each of the participants in the circle while sharing comments.

- The speaker is the only one to speak until he/she asks if there are any questions or comments.

- Raise your hand if you want to ask a clarifying question.

- This is not a time to share your own opinion.

- If you do not agree with the comments the speaker has made, ask the speaker to explain a specific thing that was said.

- If you agree with the speaker, it is not necessary to state that.

- If you agree with one thing the speaker said but you disagree with another you can say, "I agree with your opinion about _____, but I have questions about why you said _____?"

- There will be a chance to share more ideas when the leader asks for other opinions.

We can help our children get on the road to success and become resilient by overcoming the fear of asking questions and learning how to utilize questions to clear up confusion. They become more effective communicators, and by asking for help they will become empowered problem solvers.

Questions are tools to explore and investigate. They can help create meaning where none existed in the past. They can lead to action and great innovations. They can help you get to know people better and become a more engaged companion. The best questions are not what we find in Google. Great questions are ones that cause us to reflect and create something new. Our students need to know the power of their questions. With the power of the question, they will become powerful changers of our future.

CHAPTER 12

Who Taught Me to Ask?

A good question can lead to creating positive change and innovation.

This is my dad. He was the best teacher of how to be a powerful change maker by asking and answering questions. His life was driven by questions that led to his success. There was one question that resulted in his influencing the way products were sold in stores after the war in 1945.

In the 1940's, shoppers had to go to different stores to get what they needed for a week. They had to go to the grocery store for their canned and boxed goods, a produce store for fruits and vegetables, a hardware store for tools, a department store for soft goods like clothing and undergarments, a pharmacy for medicines, a convenience store for pots and pans, other necessities, and a fabric store for fabric and sewing notions.

My dad was approached by a friend who was not a problem solver when faced with challenges. Knowing that my father was a creative salesman, he approached him when he had acquired a FREE cargo container of irregular nylons. His brother-in-law gave it to him starting him off in business, but he had no idea how to sell them.

He asked my father, **"How can we sell these nylons?"**

That question was a huge game changer. My father knew he could come up with a solution, so he let the question fester for a little while. This led to more questions:

- **"How can I sell the nylons?"**

- **"Who will benefit most from them?"**

- **"What challenge do women have now that we can solve with these nylons?"**

- **"How can we make them easily accessible and affordable to women who wear nylons everyday?"**

My dad felt it was important to focus on how he could help those who might want to purchase the product. He thought, "If I can make life easier for those who wear nylons then they will purchase the product." At the time, many women were housewives, but there were many women who were still in jobs they took over when the men went off to war. In those days, many working women wore dresses and nylons even when going grocery shopping. They had two challenges. The first was that nylons ran easily and many times they only got two uses out of a pair. This was challenging since they were expensive. He knew working women were tired after a day of work, and having to go to the department store and the grocery store was a lot. The cost of the nylons wouldn't allow them to purchase more than one pair per pay period. He knew it would benefit them if they could get nylons in the grocery store where they shopped and offer them at an affordable price. This was out-of-the-box thinking.

He had already persuaded the Stater Brothers grocery chain to carry pots and pans. So, he approached them to sell the nylons. He had a convincing pitch. They agreed to provide him a paid spot, so my dad set out to build the display case. He delivered it the next week filled with affordably priced nylons. At the end of the week, he received a call from one of the brothers, "The women are buying three and four packages at a time. You don't have to rent the space, just keep the racks filled."

He met a need by asking the right questions, made it affordable, and was on his way to doing business in a way never done before.

The next year my mom gave birth to my older sister. One night she was complaining about having to constantly replace the diaper pins and the plastic pants. He knew mothers needed the same convenience as the working women, so he approached the brothers again, and they were eager to take on his new idea. He followed with more products such as teething rings and baby rattles. Before long he had one section of the aisle dedicated to over 30 items that included men and women's undergarments, sweaters, and sandals.

Since women repaired clothing in those days, and did not throw away clothing, he saw another opportunity and convinced the brothers to allow Coats and Clark, a needle and thread company, to bring in their products.

Fast forward to the 1980's and my father's concept grew into the one-stop shopping concept of stores like The Treasury, K-Mart, Gemco, and Walmart. It all started with a question.

CHAPTER 13

Why Do We Need a Paradigm Shift in Education?

As a consultant, I am continually asked, **"How can I, as a new teacher, provide the best education for their students and prepare them for the future?"** *My answer requires a simple paradigm shift in education.*

It's time to re-evaluate what was done in the past and restructure our schools to provide new experiences that are **not** connected to their tech devices or focused on repeating information. More time needs to be spent focusing on creative thinking without getting help from parents or the internet for answers.

COVID punctuated some flaws in our current methods. Focusing on passing tests was impossible when everyone was scrambling to adjust to COVID learning. It resulted in a decrease in the amount of curriculum covered and how it was delivered. The 19th century methods had worked for a long time but did not serve children well while transitioning to online learning. If teachers, parents, and children were reliant on the internet for answers, they were lost. My students weren't because their parents were completely removed from helping with schoolwork. They loved that their children were independent, especially when the parents were thrown into having to work online themselves. My eight-and nine-year-old students solved problems their own way and helped me create a learning environment online that worked for all of us. Sometimes their ideas failed, but they were used to trying and failing, so they could succeed. They knew how to try different things until they found what worked.

The recent past demonstrates why empowering children with the ability to ask questions and to independently find answers are skills necessary in the 21st century. We can't rely on past strategies when things are changing exponentially faster every day.

According to Dr. Ken Robinson, "Public education puts relentless pressure on its students to conform. Public schools were not only created in the interest of industrialism but created in the image of industrialism. In many ways, they reflect the factory culture they were designed to support." This was fine at first for a segment of the population, but it is no longer appropriate for the 21st century student.

In the 1800's, education was focused on reading, writing, and arithmetic. Not everyone went to college. Some schools' curriculum included geography.

In the 1900's a shift included the sciences. Technical classes such as auto mechanics, wood working, typing, and sewing were added to the curriculum. Later on music and art were added.

In my 65 plus years of experience in education both as a student and a teacher, I have seen many subtle paradigm shifts. The last big shift occurred when education became test driven.

Beginning in the 1980's, schools' scores were printed in newspapers. Parents were led to believe that schools were failing their children. Frustrated, and concerned parents began stepping in to help their children. They now felt a need to fill in the supposed gaps in their children's education. Without training in effective teaching practices, parents grew frustrated and began providing answers to questions when their children became frustrated with homework. It quieted the children, and the homework would come back done perfectly. This action was misleading teachers to believe they were being effective with the methods they were using during instruction.

Traditionally, students' mistakes were a guide to teachers about where students had holes in their learning. The actions on the part of the parents simply led to what they feared. Student performance and engagement in class declined.

Teachers were also seeing side effects from parental help, but they didn't make the connection. Students had no need to ask questions in class when the teacher was teaching new concepts, because they had parents at home that would help them. The minute the students got confused, they would go to Lala Land (a safe place where one does not feel dumb) because they knew their parents would provide the answers to the homework later. This created a vicious cycle. The student went home, had a fit, the parent stepped in and helped or did the work, and the teacher got the impression the instruction was just perfect for the students and moved on. As a teacher, I thought I had mastered my craft. I later discovered that was further from the truth.

As parents became more involved and did more of their children's work, teachers began to see an increase in inattentiveness in the classroom. As technology became integrated into their lives, students began to rely more on the internet to find the answers. This does not promote critical thinking and results in a generation of students who believe everything they read on the internet. Teachers who have been in the field for over 25 years have seen a steady decline in the creativity of their students.

I left public education when the push to evaluate instruction was based strictly on over testing of children. In a "teaching to a test" situation, there is no time to weave in exploring ideas with depth and complexity. There is little time to cover all the material demanded by the state, so it becomes hours of homework. There is no time to apply the knowledge students are learning. I believed the new focus of cramming in curriculum just to expose students to concepts that appeared on tests would result in a lot of educated students who didn't retain and could not apply the knowledge they stuffed into their short-term memory to pass tests.

There was also a shift in the belief that every child should go to college, and we saw the disappearance of needed technical and trade classes. Parents adopted the belief that no one could be successful without a college degree. At the time, it seemed to be correct. Over the years, more and more students were filling the colleges, some not finishing, while others received multiple degrees.

The pressure of such programs left a large population of students with huge debt, depression, anxiety, and without the jobs they were once promised if they went to the "right college." I still hear parents saying, "My child picked the college that had a reputation of getting a job out of school." I had that same guarantee in 1971, but the colleges didn't anticipate a flood of teachers for each job available. So, my guarantee of getting hired as soon as I finished student teaching resulted in me being one of 1,000 applicants for every available job. Luckily, I got a job, but thousands of my peers did not.

Students who learned information for passing tests and qualifying them to get into prestigious colleges came out of college with a belief they should earn six figures. The negative impact took time to be revealed.

C. E. Daniher

C. E. Daniher, a highly respected engineer in the aerospace field, has been instrumental in his work on projects such as space capsules, the Space Station, and shuttle orbiters. He was also a consultant on aerospace projects with other major companies. He shared an experience working with this new generation of engineers when he was asked to consult with a group of young engineers. The company's president clarified that he wanted him to deal with a "little problem" he was experiencing. He referred to his new hires as entitled new graduates from our best technical schools.

His challenge was as follows: "They graduate at the top of their class and expect to be hired at the Vice President level, but they don't know how to challenge what the computer spits out! They want top pay and feel that the company should engage their desires." **In addition,** *"If they don't get a sterling evaluation and a raise, their parents ask for a meeting to protest and advocate for them!" Can you mentor them, so they become productive engineers?'*

Daniher's first meeting with what he called 'High Pot's' (High Potential future leaders) was to orient them on a new contract in developing a plan getting to Mars and back.

His first encounter with what he referred to as 'newbies' was a shock. "The bright shiny faces were glancing up from their open laptops while a few wore earbuds which were competing with my orientation. Since they were <u>not ready to receive</u>, I focused on what they were presently doing. One was, in fact, taking notes (NO EAR BUDS) the others were emailing friends and family, playing computer games, looking at search results. They all said that this was the new way you know, multitasking."

He used his experience as KORL (NASA call sign for the Orbiter

Refrigeration System Lead for SKYLAB the world's first Space Station)
to share the difference between productive and non-productive multi-
tasking. "I agreed that multitasking was done, but within a focused
task and __not__ on unrelated tasks." They did not agree, so I stopped the
session and told the President I could not be effective with these High
Pots. He said, 'Mr. Daniher, why don't you let me have a private talk
with these youngsters then I'll call you back in.' Maybe what he said
went something like this. 'Do you want this job we hired you to do or
not?' or 'Do you know how much we are paying this guy?' I don't know
what he really said but either way, upon my return, laptops were closed,
ear buds out and we were on our way to Mars."

The second lesson they needed to learn was how to determine if the
results produced by their computers were correct and doable. "These
young engineers were able to come up with some amazing low energy
Mars' trajectories, systems definitions, schedules, and costs for doing so.
They knew it was right because that's what the computer printouts and
graphs told them. 'Why spend the extra effort checking when we could
be creating new stuff?' was their retort.

After a few hands-on examples and false starts, the answer became clear.
The projections were not correct for this project. What you gain, over
time, is the ability to "thumb" the results. Soon they got more proficient
at vetting and questioning the information and finding alternate ways
to "thumb" the answer."

He asked them important questions, "Would you recognize the answer
if it were given to you? Does it satisfy the customer's intent? Is it doable
or does it require something that doesn't exist yet? Are there alternatives
that satisfy multiple customer objectives? Is the question clear and are
there other questions that should be answered first?"

"What they learned was that not all effort is perfect, false starts can
teach you a lot and formulating what an answer should look like helps
you develop it."

Many students were trained to rely on computer generated material to do
a project. They don't realize that a new project is going to present challenges
that a program can't project since it hasn't happened before. The programs

are based on what the programmers have used for data. As my colleague, Devin Seifer, stated, "To think out of the box ideas, one must get out of the box."

1994 was a pivotal year for me. It was the year that I was able to move away from teaching to a test and begin the training I have shared earlier in this book.

I was fortunate to find a school where the founder, Dr. Laura Hathaway focused on the developmental needs of the children. She was fiercely against testing. Her passion for creative approaches to learning attracted other visionaries. One such visionary was Mary Karaba. Her Outdoor Education program helped young children's minds grow as they interacted with nature in a multi-sensory way. Along with others, I was also given the opportunity to develop a program that promoted critical thinking and training in valuable life skills that served our students long after they left the school. Tests were used merely as a tool for teaching skills, and how to productively process failure. We only did a standardized test once a year to meet the state requirements. It was a tool for us to see if we needed to revisit some concepts, or to see if there were red flags that deserved further evaluation. Our children were moving at their own rate and thriving above all schools.

As a third-grade teacher, it took four years for me to see the negative impact of shifting from a developmental program to an academic one. The children entering my classroom **from a developmental program** were excited about the new phase in their lives. Since not everyone was developmentally ready for reading or grasping math concepts at the same level, we leveled our reading and math programs. The children were very creative, because they had many opportunities for creative play.

The most prominent change in my students with the shift to an academically driven program was the presence of increased anxiety about making mistakes. They were also expressing negative feelings about their math and reading abilities. They felt inadequate at the young age of eight. When asked when they decided they weren't good at a subject, I was given the same answer, "In preschool or kindergarten." Their reasons varied, but all pointed to receiving material before their brains were developmentally ready and observing some classmates easily grasping concepts they could not.

Another glaring issue magnified by COVID was the need for a paradigm shift from just exposing children to material, to focusing on the most essential skills. Prior to COVID, educators believed that children needed exposure to all concepts each year whether they were cognitively ready for them or not. Many teachers were instructed to be on a specific lesson on a specific day even if the children were not ready to move on. Teachers were moving too fast for many children, leading them to believe they weren't capable.

During **COVID we had to slow down. Several school principals shared that this resulted in better performance on standardized tests. "We realize now, less is best," Sharon Giles, Elementary School Principal.**

A big paradigm shift has been technology influenced. I am a firm believer in mindful use of technology, but not for children's recreation. COVID proved we need to get children off devices if we want them to thrive in the 21st century. They had more opportunities to interact with nature. COVID forced our students to deal with challenges for which Google did not have the answer. They needed to learn how to play without a referee and how to occupy their quiet time without parental direction.

The true thrivers of the future will know how to access their innate abilities and apply their acquired knowledge to come up with unique solutions resulting from the questions they ask, just as my third graders did during COVID.

Bernie Marcus, founder of Home Depot, sums up what is needed in our schools to prepare children to be the entrepreneurs and inventors of the future. *"When we built Home Depot or when we became involved to support the Center for Disease Control and with veterans' causes, we had no idea where we were going to end up, and no Harvard MBA could have helped us. We were in uncharted water, and there was nobody to tell us what was right or wrong. But we had plenty of common sense, something I inherited from my mother. She taught me to look at a situation critically to really understand the issue, to trust my own judgment, but not be so stubborn that I wouldn't listen to other viewpoints and take advice. She also taught me not to sit around and wait for others to fix things."*

CHAPTER 14

How Did Asking Play
Out in My Life?

Questions I've Asked

I've been asking questions since I was a little girl. If I asked my dad why I had to do the dishes, he would reply with, "What do you think?" So, I naturally was a question asker.

My students in the 70's had longer attention spans than the children of the 80's. I couldn't understand what was causing this change. My peers thought it was the influence of Sesame Street, but I wasn't sure that was true. I went to bed one night wondering, "Why are children so inattentive today?"

Suzy Was the Answer

*The answer came the next day in my second-grade classroom. In the 1980's, I thought I was getting good at my craft with 8+ years under my belt. Until Suzy, a 6-year-old in my second-grade class demonstrated why students were becoming so distractible. It showed that I wasn't as good a teacher as I thought. I told her that if she did not concentrate during my explanation of the homework, she wouldn't know how to do it. Her reply while she played with the toys in her desk was, **"That's okay, my mommy will help me."***

When I called her mother, the mother shared how horrible her evenings were after a hard day at work. Her daughter would throw fits every night because she didn't know how to do the homework. Suzy's mother said, "I was so tired, I just gave her the answers. I didn't know how to teach her." I gave Suzy's mother permission to stop helping, and I told Suzy that her mom and I agreed that it would be better for her to come to me with questions if she wasn't clear about homework. This changed Suzy's attention in class and saved her mom from Suzy's nightly tantrums.

I knew I needed to stop the vicious cycle by encouraging my students to ask lots of questions. That's when I discovered the *fear* children had about asking questions and their **lack of knowledge of how to ask for help.** I began offering parent seminars on how to shift from being a homework teacher to a homework coach. I offered them strategies to use with homework challenges and empower their children to get the support they needed without parental help.

When the state began placing more emphasis on test taking, I knew public education would not allow me to help students develop the skills they needed to be successful in life. There was no time to allow children to fail successfully. When seeking a new teaching job I asked, **"Where am I going to find a school that won't focus on testing?"**

A very dear friend had taken a part-time job at The Pegasus School in Huntington Beach. She suggested I apply for a teaching job. After meeting the founder, Dr. Laura Hathaway, I knew **her school was the answer to my question.**

She encouraged me to implement the programs I felt our students needed to grow into productive independent citizens who would make a positive impact on our world.

Parents were very high achievers and wanted their children to be the best. They often stepped in to help with homework, so teachers were assuming the instruction they gave that day was effective. The question I asked was, **"How can I convince the parents to trust me?"** I knew they were doing what they thought was best, but they had no idea of the negative consequences.

I'd heard teachers say, "Parents should _____" about any challenges the children were having. I didn't want to "should" on them. I believed deeply that if they knew a different way that would get them the results they wanted, they would do it. I had to be able to be the only teacher for the children, and I asked the parents to stop helping. This frustrated parents who felt their children couldn't succeed without their help. My next question was, *"How can I ease the stress of these parents?"*

The next morning, I awoke thinking, *"In order to remove parents from helping at home, I had to help change their role to that of a coach so they could support what I was doing in the classroom and feel like they were doing something."*

But that prompted another question, *"How can I convince them it will help them get what they want by making this change?"*

The answer came, "You need to teach them some strategies to support what you are doing in the class. Make them feel they are part of a team effort."

I saw the need for parent seminars that showed them how they could help support my work in the classroom and relieve their fears. I offered neuroscience supporting the value of their children asking questions in class, which shifted their role to that of a coach instead of their home teacher.

I also had to make sure the children knew their role on the team. For them to be successful, they needed to be responsible for letting me know what they didn't know about the homework. I asked them to ask me every question they could think of about their homework before leaving for the day, so they would be prepared.

What I discovered was the students didn't know how to ask for support or were afraid to do so. My next question was, *"How can I get them over their fear?"*

As I described earlier in this book, the answer came at a *Learning and Brain* conference. I needed to share neuroscience with my students. I realized using basic neuroscience principles would help my students feel comfortable asking questions.

The parents of my students were asking me to write all my strategies down because they wanted to refer to them. My goal was to help them shift from a homework teacher to a homework coach, so I reluctantly took on the task.

Some people suggested I wasn't qualified to write the book, because they claimed I didn't have data to prove my theories. I didn't need someone else's data to prove what I already knew. I had two daughters who were using the strategies successfully and former students were thriving because of them. I was told, "You need a master's degree to get credibility." Because of the prodding of parents who were amazed at how their children were transformed, any doubts I had were quickly overcome.

The big game changer came when I first doubted myself. I was awakened abruptly at 2:00 am by a booming male voice, *"This is not your choice. This is your responsibility. That's why you are here!"* I haven't shared this in the past, but it was the push I needed to work through the challenges writing presented to me.

I was almost finished writing my parent's guide when I thought, *"What do I do next?"*

I spent time in Kona during the summer of late 1990's, where I met Mark Victor Hansen and his family at a bay in Kona Estates. I asked him about the necklace he was wearing. He shared that it was a gift from the Nikken company for motivating their employees. He had just purchased the property that was right in front of us. After a nice conversation, he asked me what I was doing on the island. "I'm babysitting my great-niece and finishing up my book at the same time." The answer came when he randomly said, *"Self-Publish."* I confessed to him that I didn't have a master's degree. He adamantly reassured me that my experience in the classroom for 27 years was the best education. At the time, I didn't know about his Chicken Soup book series. He shared his own story about publishing the first book, which helped me seek a publishing company.

I knew nothing about self-publishing, so I asked, *"How the heck does anyone do that?"*

I was discussing this with my mother when she shared the answer I was looking for. *"I just saw an ad for a self-publishing seminar at the Claim Jumper this Thursday."* I was already booked for a parent seminar, so my mother went for me. With the information she discovered, I was on the road to my first book becoming a reality.

Because of the excitement of my students' parents, I fully believed the 2500 copies I published would fly off the shelf. After selling the first 100 to present and former parents, I was left with 2,400 copies that didn't sell. The next question was, *"How am I going to get these books to people who need them?"* I was in a quandary and wondering if I really was on the right track. If it wasn't for the encouragement of the parents who benefitted from the program, I would have done nothing further with the remaining books.

The school librarian's suggestion was the answer, *"**Apply to speak at the California Association for the Gifted Conference where you can sell your books. After you speak, everyone will want to buy them.**"*

I presented two sessions to a ballroom full of teachers and parents. *"Homework Solutions for Weary Parents and Teachers"* proved to be a hot topic. The ballrooms were full of people sitting against the walls, in the aisle and spilling out into the hallway. At the end of the sessions, I had sold 800 books and was offered several speaking engagements. My business began to grow. Within a year all the books were gone, and I tackled the job of a second edition.

After printing my second edition and a teacher's guide, I needed help promoting them. I was listening to Tony Robbins' tapes while walking on a deserted beach one December. Robbins' lecture focused on how to ask. No sooner did I ask, *"**Who do I need to connect with for help with this book?**"* Within a few minutes of asking my question, a woman in a red dress, the only one present on the beach at the time, came walking down the strand. She told me she had this sudden urge to walk along the beach instead of going through the parking lot. She was my former principal who had connections with an educational consultant who was able to open doors to more seminars and work on more books.

In 2002, I came across the book *Write it Down, Make it Happen: Knowing What You Want – and Getting it!* by Henriette Anne Klauser which resulted in my attracting companions in record speed. I began using it to manifest jobs for my business. While doing a seminar in Kona for the students at Konawaena Elementary School, I got some snorkeling time. I was newly divorced and wondered, *"**How can I be here for the whole summer? I need the island to help me get a fresh new start. Where can I stay?**"* That evening, I was visiting with some friends. They had just bought a dog. I offered to watch the dog in California, so it wouldn't have to be in quarantine for six months. The next day they offered me their house in Kona for the summer.

My question was answered. Seek and ye shall find!

Fast forward two years from my first meeting with Mark Victor Hansen.

I had sold out of the first run of 2,500 copies and was already selling the second edition. I wrote two more books. Questions others asked inspired me to create parent, teacher, and student seminars, and write three additional books.

Parents continually asked me, "When this happens, what can I tell my child to say and do?" Not having been taught skills themselves, they were at a loss about how to empower their children without doing the work for them. I had completed the routine for the book _Homework Solutions for Weary Students and Their Parents_, but because it was a collection of random situations it posed an organizational challenge. I went to bed asking, *"How am I going to organize this?"*

The next morning, I gave my students a well-earned break from the daily routine and allowed them to read anything they wanted for 20 minutes. One boy came up to me and said, "Ms. O., I am having a challenge. This book gives me two ways to go with the story and I don't know which one." I was looking for a way to organize my new book, and the answer came in the strangest way. His question led me to create my table of contents to read, "If this is your problem, turn to page _____."

I hugged him and answered his question, "If you can't choose, what can you do?" He answered his own question, "I could read it both ways and see which one I like best."

Jack Canfield was a featured speaker at a conference where I presented. I never dreamt four years earlier that I would be speaking at conferences and doing seminars all over the United States. I began setting my sights on doing what Canfield was doing. At the end of his presentation, *I asked him if I could take him out to lunch.* I asked for advice on how to become a keynote Speaker. He asked me, "Why aren't you?" I wasn't sure, so he agreed to meet with me. He suggested I put my books on Amazon. It wasn't long before I was a featured speaker at a national educational conference in Palm Springs.

When I was preparing to speak to him, I did some research and saw an ad for a Mega Book Selling seminar by Mark Victor Hansen. I had sold enough books at the last conference to pay for the three-day seminar. The seminar was powerful in learning to say "yes" to everyone who asked me if

I did different types of seminars. Their questions resulted in expanding my seminar topics and increasing my publications.

I took his advice and pre-sold enough books to pay for the printing of 11,000 copies of two of my books.

Life has a way of throwing us curve balls and my business had to go on hold while I shared the care of my 90-year-old mother. Then COVID changed the way we did business. I have discovered that I do best when I am with a live audience. I adapted to the Zoom method and did some seminars on-line on a variety of topics designed to ease the stress for parents and teachers. I used this isolated time to take several courses on-line, but still felt like I was floundering Everything I read, and the classes I took during COVID prompted me to ask yet another question.

I have developed several programs and I needed to know what to focus on, so I asked, **"I want direction. Where do I focus my energy?"** *The next morning, when I opened Facebook, there was Mark Victor Hansen advertising his new book,* <u>Ask-The Bridge From Dreams to Reality</u>. *It validated that what I have been teaching my students was exactly where I should place my focus and what is needed to meet the demands of the 21st century student.* One of the last points made in his book was that we need to learn to ask like children. I had spent the last 40 years teaching children and parents how to overcome their fears and do just that.

I am not done asking. Even last night, I asked this question, after having read and re-edited my book for the 20th time, *"What is missing that is very important?"*

I dreamt the answer. I was telling my students what had happened to me the night before and I was wondering what questions they would come up with to help me solve my problem.

I awoke with the need to reread my work and add in how important it was for parents and teachers to model tackling real-life problems by asking questions.

My questions get more specific and continue to be answered in the most unusual and unexpected ways. Teaching children how to do the same needs to begin at a young age to prepare them for the unpredictable challenges they will face in the 21st century.

CHAPTER 15

What Are the Author's Credentials?

Victoria Olivadoti has passionately empowered students for over 52 years to advocate for themselves and be the question askers of the future in school and their careers. Her unique approach to learning has earned her the honor of being asked to speak at national conferences, for school districts, and Parent Teacher Organizations. Her contributions have been recognized by Who's Who in Education, Who's Who of Young Educators, and Who's Who in America. Victoria recently retired after 50 years in the classroom. It wasn't long before she found herself with a new career shift into the Homeschooling field. She currently works with students on-line and continues to provide her support to teachers and parents through her books and seminars.

She has taught children from kindergarten to eighth grade, while also providing courses through the Univsersity of California at Irvine. Her strategies have empowered high school and college students as well. Teachers nation-wide have attended her seminars in which she shares her powerful strategies. Victoria is a columnist for local newspapers, a blogger, and has authored several books. As technology has evolved, she saw a need to teach skills so students could learn how to control their computer devices, so the devices didn't control them. She also created a program that educates and empowers students to effectively utilize their time, manage their papers, effectively communicate, and advocate for themselves. Her students are powerful creators and see challenges as opportunities for positive change. Victoria's methods have created paradigm shifts in how others view their responsibility as parents and teachers. She has mentored hundreds of parents from "helicopter" and "lawnmower" parents into becoming powerful mentors of their children. She helped teachers gain true sustained engagement in their classrooms by making simple changes in the methods they used.

What Do Others Say About This Book?

"Ms. Olivadoti has helped hundreds of students achieve success with her methods. Students and parents know that the system works. Students learn that effort and persistence not only bring academic success, it brings confidence and a "can do" attitude - an invaluable skill that is theirs forever."

–Dr. Laura K. Hathaway, Ed. D.,
Founder, The Pegasus School

"**Questioning** by Victoria Olivadoti should be provided to every student. It contains essential information based on experience and neuroscience and how to be successful in the 21st-century. As an educator, and more importantly, as a parent, I have witnessed the power of these techniques in helping the lives of children including my own."

–Terry Burnham,
Associate Professor Chapman University,
Co-author The Secret Syllabus

"I still use the strategies Ms.O. taught me in third grade. All through middle school and high school I used office hours. I was once afraid to talk to the teachers, but now I really enjoy it. I know I need visual support in the classroom. Once I asked for it, the teacher did it the whole year."

–Samual Abrams, Graduate Student NYU

"I really believed that asking parents to help their children with homework helped reduce their stress. After reading your first book, my son said, 'You thought I didn't know that you were fixing my homework after I went to bed. I never felt my best was good enough for you. I had to get counseling in college to deal with it.' After reading your book and seeing your students thrive, I'm glad I could do the same for my students."

–T. Frantis, Third Grade Teacher

"Ms. O, I wish people my age had been in your class. They don't know how to read between the lines. They don't question anything. I'm grateful that I learned at eight years old."

—Rachel Wilson, 30-year-old

"Your book made me think about things I never did before. I will use these strategies in the future."

—Kavan Dang, 7th Grader at The Pegasus School in Huntington Beach

"I wish I had known this when I was little. I think I would have made different choices. The students at my school need to read this."

—Kitty Rede, Administrative Secretary, Laguna Beach High School

There is so much worthwhile information in this book. It should be required reading for all teachers, parents, and students. Honestly, all the strategies and skills are great for adults in everyday life. Victoria is a gifted instructor of life coping skills!

—James Corlett, Owner AMI Pet Innovator Imaging

This is a book for everyone, students, parents, grandparents, and everyone else who wants to learn why and how to question. This book is so needed right now to help everyone navigate and ask questions. You could not have written it at a better time. The examples and words you used to help your students (and everyone) ask the right questions are so important. This book offers parents and teachers tools to help children face their fears and ask questions for themselves.

—Karen Wells, Former Parent

I appreciated how Victoria Olivadoti approached "Questioning" from several angles. She clarifies why asking questions is an essential 21st Century skill for students? She offers the value of knowing where and when to ask questions, and how neuroscience validates asking questions?

Ms. Olivadoti's research displayed rich use of resources and her 50+years of experience to support the value of questioning.

It is my belief that educators would reap true benefits by learning from her solid work.

—Minard Duncan, Retired Principal,
Fullerton School District

Contact Victoria Olivadoti at Victoria@homeworkdoc.com for information regarding her books, seminars, and private coaching.

Seminars

Victoria is available for Conference Seminars, Teacher In-services, and Student-focused Classroom Sessions

Seminar #1: Restructuring lessons to support virtual learning©
(1 Hour)

The classic lesson plans of the past are not effective methods for on-screen delivery of material. Internet lessons drain children of mental energy, which results in less absorption of concepts. Internet transmission is challenging, because of breaks in connection. The ability of children to ask questions is impacted by the teacher's attention to the delivery of lessons. Using the methods I personally used in on-line learning, I was able to shorten lessons, while increasing the students' understanding. The methods I will share, reduce the time needed for direct instruction, pinpoint student's unique needs, while offering differentiated instruction that results in better retention of the concepts taught.

Seminar #2: Technology Management©
(1.5 Hour)

Technology is designed to control us and there needs to be mindful use of technology. Learning at a young age how to control technology and break the addiction-loop cycle is crucial to its effective use and improved time management. Learning which platforms match one's learning style, while learning to manage the invasiveness of technology, are vital skills. Helping students understand how AI is created and the pitfalls of using it without questioning it are crucial now that AI is attached to everything we see, do, and say when using devices.

Seminar #3: Secrets About the Brain Every Student Needs to Know©
(1.5)

Understanding how the brain works is very helpful in all learning models. The COVID learning experience made it even more important to understand the brain's stress response to on-line learning and the challenges faced by the unknown. Students pre-trained in managing their own stress response adapted to the COVID learning model easily. With the neuroscience knowledge presented in this seminar, students will learn how to work smarter instead of harder. The results will be effective stress management, focused attention, effective access of prior knowledge, and maximized performance.

Seminar #4: Seven Secrets About Teachers Every Student Needs to Know©
(1.5 Hours)

This session offers a different perspective of the teacher's role in the classroom and introduces a new definition of the role of the student. This approach helps teachers establish a safe and friendly learning environment.

- Creates a more human and approachable image of teachers
- Lays the foundation for establishing effective student-teacher relationships
- Emphasizes the need for clear communication between teacher and student before, during, and after instruction
- Defines clear observable behaviors of successful students

Seminar #5 Time Management and Backwards Planning: The Cure For Procrastination©.
(2 Hours)

Learning to independently manage one's own time is a life skill every student needs to be successful in careers that have yet to be determined.

This session presents an approach teachers can share with their students to solve the challenges students experience with daily, weekly, and long-term independent as well as group projects. The process of teaching time management is very specific, easy to implement, and quickly helps students manage and meet deadlines independently. This approach to long-term projects has proven successful for over forty years and has helped former students experience success throughout their educational journey and into their professional lives. It is time-tested and proven to be one of the most valuable tools for students to have in their toolboxes. It also gives meaning to our mantra "Proper Prior Planning Prevents Poor Performances." Through this unique approach to planning, students gain the following:

- The value of clarifying expectations as soon as a project is assigned

- How to create due dates for different parts of a project based on their unique time needs

- How to allow for the needs of all members in a group project

- How to determine the steps needed to complete a project independently

- How to account for unexpected time needed to complete a project

- How to ask for and receive guidance during the execution of projects

Seminar #6: Overcoming the Fear of Asking Questions and Making Mistakes
(1.5 Hours)

Self-advocacy is essential in a traditional classroom and an online learning model. The fear of asking questions can paralyze students and negatively impact their performance. Relying on parent support is ineffective. Supporting the unique learning styles of their students requires teachers to encourage questioning. For that to happen, students need to overcome their fear of asking for clarification and support. This seminar helps teachers learn how to coach students to ask questions effectively and overcome that fear. When students ask questions, they want an answer. This results in full engagement. The students, who overcame the fear and were trained to ask questions, adapted beautifully to the on-line learning model during COVID. They not only survived, but they also thrived.

Seminar #7: How to Engage Parents as a Team Member
(1.5)

Parents are not equipped to be in-home teachers. On-line learning has pushed them over the edge. Most parents report homework has created havoc in the homes. They hate being dragged into the role of teacher when they are not equipped for the job or because they must work to put food on the table. Understanding the unique needs and fears of parents, along with their perception of what a good education entails, is important for all teachers.

Educators also need to create a team approach to this learning model. If parents already knew how to do things differently, they would. This session will offer educators the tools they need to create a team relationship with the parents, students, and themselves.

Seminar #8: Digital Dangers: A Parent's Guide to Proper Prior Preparation to Prevent Device Users
(1.5 Hours)

This seminar addresses the skills children need to be prepared to use devices safely. Parents warn their children about "Stranger Danger" in parks but are not aware of the many dangers lurking behind their children's computer, tablet or mobile phone screens. With the invention of AI, we don't know how strangers will manifest nor what they will look like. This seminar will help parents determine their children's readiness for device use and offer important information to prepare their children if they choose to hand them a device. Removing devices is not the answer. It is a part of our future, but children need to be prepared to use them. This seminar offers an assessment to use in determining readiness, just like a student is alerted to the dangers on the road before they receive a driver's learning permit.

Books by Victoria

Backwards Planning Workbook-Cure For Procrastination (Ebook)

In an age of digital distraction, it is becoming increasingly more important to train students how to manage distractions and plan for projects. This method has proven effective for students nationwide.

https://homeworkdoc.com/product/backwards-planning-workbook-cure-for-procrasti- nation/

Homework Solutions: A Parent's Guide-The Missing Link Between Home and School

The "Homework Solutions" series presents a creative and innovative approach to help students, parents and teachers manage the "hornet's nest" of homework. This book offers a new perspective of the parent's role in their children's learning. It offers strategies that will guarantee their children's success.

"This book is an excellent resource for anyone interested in the success of students."

—Kathy Coleman, MA Inclusion Coordinator Facilitatory/Mentor - "Schools Attuned"

https://homeworkdoc.com/product/homework-solutions-a-parents-guide-the-missing- link-between-home-and-school/

Homework Solutions for Weary Students and Their Parents

Homework Solutions for Weary Students and Their Parents helps the parenting struggling with assisting with homework. Children need to communicate confusion with their teachers. Knowing what to say and how to express confusion are powerful skills that will guarantee children get their needs met in the classroom.

https://www.amazon.com/Homework-Solutions-Weary-Students-Parents/dp/ 0971330204/ref=sr_1_2?crid=KQ8S6RJJUMLT&keywords=victoria+ Olivadoti&qid=1705271398&sprefix=victo-ria+olivadoti%2Caps%2C183 &sr=8-2

"This book is a goldmine of techniques from which students ages eight to eighteen can benefit."

–John Sullivan Assistant Head
The Pegasus School

Homework Solutions: A Teacher's Guide-The Missing Link Between Home and School

The "Teacher's Guide" is an innovative classroom application of homework. Teachers love this new approach because it creates 100% engagement, while organically differentiating instruction. Our children are living in a world much different from the ones most teachers experienced. It calls for a different way to create engagement. If children think their homework will be easier if they ask questions about their assignment before they leave class, they have a reason to listen.

"When I shifted to this method, I was shocked at the number of questions that were prompted by what I thought was a simple assignment."

–Danielle Field, Third Grade Teacher

https://www.amazon.com/Homework-Solutions-Teachers-Guide-Missing/dp/0971330247/ref=sr_1_1

Memory Joggers- Multiplication. - Division

"Memory Joggers" helps multiplication and division number sentences formed in the left side of the brain be visualized in the right side of the brain. Since our brains create memories by forming pictures, Memory Joggers help the memorization of the fact families. Rote memory is quickly lost and requires more practice than using the Memory Joggers method. Each fact family is depicted in a picture that is accompanied by a story and a script parents and teachers can use to help build recall. This system was a total game changer for my students who were artistic and had challenges memorizing facts.

https://homeworkdoc.com/product/memory-joggers-multiplicaton-division/

"I still use your strategies when working for Van's Shoes. I hated math and never felt capable until I saw math your way. Now I do math as a career."

—Erin Theodora, Financial Advisor for

Van's Shoes, My Former Third Grade Math Student

Set Your Intentions: A Planning Process Guarantees Success (Ebook)

Being successful is easy with a little pre-planning. The best place to guarantee a successful school year is before school begins. This handbook provides students with a guide for doing just that. Following the simple steps in this workbook will get students off to the right start.

https://homeworkdoc.com/product/set-your-intentions-workbook/

Sleep in Children: A Practical Guide for Parents by Melissa Olivadoti, Ph.D.

Sleep deprivation and poor sleep quality are an epidemic in America. In a 24-hour a day, 7 days a week society, we don't value sleep as something necessary for good health and well-being, and children are paying the price. Sending students to school sleepy is like sending them to school without their paper and pencils. They don't have the tools necessary to learn the skills they are being taught at school and are at a higher risk for getting sick, becoming overweight, and more likely to exhibit behavioral problems.

https://homeworkdoc.com/product/sleep-in-children-a-practical-guide-for-parents/

Top Secret: What Students Need to Know About Teachers: Key to Student Success (Ebook)

This book offers powerful insights into teachers. With this information, students can take charge of their education and learn how to get their needs met from any teacher and future bosses.

"This information is helpful. It helps me see my teachers differently and makes it more comfortable to ask questions in class. It's been helpful learning how I learn, and I have seen how much easier school became when I began sharing this information with my teachers."

—Madison Gorsage, Former Third Grade Student, Senior at UC San Luis Obispo

10 Digital Dangers: Proper Prior Preparation Protects Users (Ebook)©

Parents have warned their children of "Stranger Danger" in parks, but they are unaware of the dangers that lie behind their children's screens. Knowing what lurks will help parents make decisions about how and when they allow their children to use their devices. This book helps parents understand how to make mindful use of them. It teaches children about what to be alert to and how to manage their own devices. This is what will protect them in a world where technology is changing at an exponential rate. New challenges arise daily, and knowing what to look for will protect our children.

"I was not aware of the bullying on-line until my child experienced it. I wish I had known that the educational game she was playing would cause her so much mental stress, which resulted in her developing a severe case of depression."

—Mrs. Pharm, Parent of a Third Grade Student

"My son became so addicted to his device, I had to remove it from the house. His teacher called to share his odd behavior. He was rocking on the floor because of my cutting him off cold turkey from his device."

—Mr. Chin, Father of a Third Grade Student

https://homeworkdoc.com/product/users-beware-stranger-danger-lurks-behind-chil-drens-screens-copy/

Coaching by Victoria

Academic Success Coaching

Private one-on-one coaching for students from ages 8 to college. Students discover their unique needs and how to be powerful self-advocates. Students learn the value of using office hours and the importance of their questions, so they receive personalized instruction.

Device Management Coaching

Learn the neuroscience principles that cause addiction to technology devices and how to control devices instead of allowing devices to control them. Students learn appropriate and inappropriate times to use devices. With this information, they become powerful time managers. Students gain knowledge of the advertising and propaganda techniques used to target young consumers. No longer fall prey to those who act like a friend online to win their confidence and influence them to make purchases.

Time Management Coaching

Geared for students who have not been taught how to manage their time, because parents and teachers have always managed it for them. Learn strategies designed to meet their unique needs and the way they view time.

"When I did my scheduling, I realized I didn't have time to cuddle with my little sister in the morning unless I get up earlier."

–Eli, Third Grader

'The first time I did scheduling with Ms. O I could see I didn't have any time to do video games after school."

–Michael, Third Grader

"I loved the idea about turning in my work a week early. It makes me feel so much better. Now I'm not late and the teacher shows me how to improve my paper before I get a grade."

–Joshua Tompsin, High School Freshman

New Teacher Coaching

Many young teachers question their ability to teach after meeting a class of students that do not respond to strategies taught in teacher training methods' courses. They quickly see that what they were taught in college doesn't always apply in the classroom. New teachers and teachers who want personal support can share the experiences they are facing for which they don't have solutions and get solutions to them.

I Hate Math

Many students form a negative feeling about their capabilities with math. Much of this attitude is a result of students receiving instruction in concepts for which they are not developmentally ready to understand. Some are taught in a very linear way, which supports left brain learners, but leaves right brain learners in the dust. Through this different approach to math, students gain confidence, and learn how to ask for support in a way that gets their needs met from any teacher. They are in the driver's seat. Victoria loves hearing "I used to hate math, but now I love it!"

Contact Victoria Olivadoti at Victoria@homeworkdoc.com for information regarding her books, seminars, and private coaching.

Acknowledgments

This book would never have reached the printer without the help and encouragement of many wonderful individuals. I was blessed to have two parents who challenged me to find my own answers. My father was innovative and mentored by example. My mother taught me "touch things once" and how to manage time. I am fortunate to have given birth to two beautiful intelligent daughters, Shelly Olivadoti-Santoro and Melissa Olivadoti. They have filled my life with joy and challenged me to find the best way to support them as they grew. To my two grandchildren, Alexandra and Jackson for offering their input. I am certain I have learned more from them than they have learned from me.

Growing up with siblings has taught me to see multiple perspectives and understand how important it is to ask clarifying questions instead of jumping to conclusions. To every student who crossed the threshold of my classrooms over the years, I thank you for agreeing to be my "Lab Rats" (lovingly coined by my former student Danny Diamond). You were willing to trust me and keep trying until you found what you needed to be successful. I am so proud of you and appreciate the continued connection we have years after you left my room.

Karen Wells, a parent of one of my third graders, encouraged me to write my first book. Anne Morris, also a former parent, insisted on being the graphic artist for my third book, _Homework Solutions: A Parent's Guide the Missing Link Between Home and School._

Linda Brodsky Abbit's skilled editing was essential in guaranteeing my punctuation and sentence structure were correct. Tammy Lucier's keen eye for details and her insights were invaluable. I'm grateful to Jackie Pensa, Marcia Schultz, Kitty Rede, Jim Corlett, Kavan Dang, Heather Fugiyama, Tammy Lucier, and Minard Ducan who willingly read and reread multiple versions and offered honest critiques and edits.

I am greatly appreciative of the expert organizational support of Deanna Mahoney, who patiently edited early versions and offered continued encouragement throughout the whole process. I have to thank those who challenged my ideas and helped me see a different perspective. I must thank my colleague and highly respected friend, Nancy Wilder, for the hours we share deliberating the needs of our students. Her Den of Scholars provided the breeding ground for many of the ideas found in my work.

To the former parents who continue to praise the program that helped their children grow into successful adults.

To Charles Daniher for sharing his knowledge and experience, and Charles Levin for his wisdom.

Works Cited

Allen, James, *As a Man Thinketh,* Mount Vernon, N.Y.: Peter Pauper Press, 1951

Allen, Jeffrey, *Duality,* Mindvalley Seminar, https://home.mindvalley.com/quests/en/duality, 2023

Association for Psychological Science. "Curiosity is critical to academic performance." ScienceDaily. ScienceDaily, 28 October 2011. <www.sciencedaily.com/releases/2011/10/111027150211.htm>.

Bodenburg, Dorothy, *Overachieving Parents, Underachieving Children,* Lowell House Publishing, Los Angeles, CA, 1992

Canfield, Jack & Hansen, Mark Victor, *Aladdin Factor*, Berkley Books, New York, 1999

Carnegie, Dale, How to Win Friends and Influence People, Simon & Schuster, New York, 2009

Covey, S., *The Seven Habits of Highly Effective People*, Simon & Schuster, New York, 1989

Dyer, Wayne M.D., *Being in Balanc*e, Hay House Inc, Carlsbad, CA, 2006

Dyer, Wayne M.D., *Getting in the Gap*, Hay House, Inc., Carlsbad, CA, 2010

Dyer, Wayne M.D., *Meditation for Manifesting*, Hay House, Inc., Carlsbad, CA, 1998

Grandin, Temple, *Calling All Minds-How to Think and Create Like and Inventor,* Phlomenal Books, New York, 2018

Hansen, Mark Victor and Crystal, *Ask: The Bridge From Dreams to Reality*, Post Hill Press, _____ 2020

Hill, Napoleon, *Think and Grow Rich.* Jeremy P Tarcher, _NY,NY___, 2007

Kwik, Jim, *Super Brain,* https://youtu.be/KVm5QuXSxxAN

Lahsen, Vishen, *Six Phase Meditation,* Mindvalley Seminars, https://youtu.be/oeQfRtiY-ZM, 2020

Marcus, Bernie, *Kick Up Some Dust,* Marcus Foundation, Harper Collins Publishers, New York, 2022

Mlodinow, Leonard, *Elastic: Flexible Thinking in a Time of Change.* First edition, Pantheon Books, New York, 2018

Peale, Dr. Norman Vincent. *The Power of Positive Thinking.* Cedar Books, 1990

Radvansky, Gabriel, *Walking Through Doorways Causes Forgetting, New Research Shows,* Notre Dame News, https://news.nd.edu/news/walking-through-doorways-causes-forgetting-new-research-shows/, 2011

Robbins, Anthony, *Awaken the Giant Within: How to Take Immediate Control of Your Mental, Emotional, Physical & Financial Destiny,* https://books.google.com/books/about/Awaken_the_Giant_Within.html?id=eoOkPwAACAAJ, Simon & Schuster, 1992

Robinson, Ken, Ph.D., *The Element,* Penguin Group, New York, 2009

Sousa, David A., *How the Brain Learns, 4th edition,* Hawker Brownlow Education, 2006

AI Revolution; David Byrne, 60 Minutes Interview, https://youtu.be/YTSUa-7Ym5M, 2023

www.ingramcontent.com/pod-product-compliance
Lightning Source LLC
Chambersburg PA
CBHW070755120626
46557CB00002B/611